Selected Applications of
Modern FT-IR Techniques

Selected Applications of Modern FT-IR Techniques

Koichi Nishikida   *Perkin-Elmer Corporation*

Etsuo Nishio   *Nicolet Japan Corporation*

Robert W. Hannah   *Perkin-Elmer Corporation*

CRC Press
Taylor & Francis Group
Boca Raton London New York

CRC Press is an imprint of the
Taylor & Francis Group, an **informa** business

Contents

Part II.   Selected Applications of FT-IR Spectroscopy

3.   Analysis of Film and Film-like Samples   61

Preface

This volume is intended to show beginners in modern Fourier Transform-Infrared (IR) analysis which technique of infrared analysis should be selected and how to use it to obtain certain information from the most common samples brought into research and analytical laboratories in production industries such as the polymer, semiconductor, and pharmaceutical industries. As the present authors were involved in many discussions, demonstrations, and collaborations regarding infrared analysis with a variety of Fourier Transform-IR prospects and users, they realized the need for a guidebook for beginners in infrared analysis. The authors had accumulated experience in a wide range of infrared applications through day-to-day work, and decided to write such a guidebook for beginners.

The original version was published in Japanese by two (K.N. and E.N.) of the authors in 1990 as a manual for laboratory technicians and analytical chemists who had newly started working with the Fourier Transform-IR system. Because K.N. had published another volume on infrared analysis with Dr Reikichi Iwamoto, the original version, which served as the data book for the present monograph, had only simplified explanations of the instrument and accessories. When K.N. was transferred from the sales office in Japan to the Infrared Product Department of the Perkin-Elmer Corporation in the U.S.A. in 1991, we decided to translate the original version into English, fortifying Part I with more theoretical and practical descriptions of infrared techniques. In addition, we added chapters on infrared microspectroscopy and chemometrics, because those two subjects have become more important.

This volume is divided into two parts. The first section of Part I describes the hardware of Fourier Transform-Infrared spectrometers and offers guidelines on how to select instrument settings, and Section 2 of Part I is concerned with theoretical and practical aspects of infrared techniques such as IR-microscopy, ATR, diffuse reflection, specular reflection, photoacoustic detection, and emission spectroscopy as well as data handling techniques. Part II consists of about 90 examples of analyses using these infrared techniques. For some samples, we selected multiple techniques to obtain different information from the same sample. In some applications, we have shown some spectral artifacts due to improper use of the techniques in order to show the importance of understanding the optical processes of such infrared techniques. In some applications, for example, chemometric applications which usually require a large numbers of standards, we have shown very simplified experiments to make the application more understandable.

Although many applications written in this book are more or less well established methods to characterize samples and have been published elsewhere by many pioneers in Fourier Transform-IR spectroscopy and by pioneers even before the modern Fourier Transform-IR spectrometers became available, we have included some previously unpublished new technical data we have developed. For instance, migration of the lubricant of hard disk media is sometimes discussed but not fully understood. The data shown in this book will prove that the lubricant migrates due to centrifugal force. A Gel Permeation Chromatography-IR technique, which is one of the operation modes of Liquid Chromatography-IR technique, is not well documented so far. However, it has excellent applicability in polymer characterizations, as shown by some of the applications in this book.

Readers may be surprised by some negative comments regarding some techniques. We have included examples of improper use of certain techniques to certain samples, for example, diffuse reflection method applied to the analysis of thin layer chromatography spots. All of the experiments were performed by the authors, except for a few cases where the data were obtained from colleagues. The reader can follow the method described in this book to obtain spectra at least as good as those shown, even on samples which are not explicitly used in this text. However, in order to understand the applications, the reader should read the original papers cited.

We wish to thank Dr Reikichi Iwamoto, the Director-General of the Government Industrial Research Institute, Osaka Agency of Industrial Science and Technology of MITI, for allowing us to summarize a large portion of the chapter on ATR technique written in the monograph he published with K.N.

We are grateful to those who donated samples; the late Dr Nobukatsu Fujino of Kyushu Denshi Kinzoku Co. Ltd., Dr Michiya Fujiki of Nippon Telefax and Telephone Co. Ltd., Mr Kazuhiro Yamada of Yokohama Tires and Rubber Co. Ltd., and Mr Eiji Hosoda of Tokyo Ink Co. Ltd. Mr Senya Inoue and Mr Jyunji Furuhashi of Kanto Kagaku Co. Ltd. gave us permission to use unpublished data. Dr J. McClelland of MTEC Corp. contributed interesting Photoacoustic data on fibers and degraded polymer film. Drs Richard Spragg, Ruppert Aries, and D. Lidiard of Perkin-Elmer Ltd., U.K., provided valuable data on chemometrics and Drs David Schiering, and Brian McGrattan and Gregory McClure of Perkin-Elmer Corp., U.S.A., provided data on microscopy and hyphenated techniques, respectively. We are also grateful to Mr Mitsuhiko Morimoto, Mr Noriyasu Niiya, Miss Sakae Niimi, and Miss Takako Tokura for their technical assistance.

While the translation and revision of the book was in progress, R.W.H. retired from the Perkin-Elmer Corporation and E.N. left at Perkin-Elmer Japan and moved to Nicolet-Japan. As a result publication was delayed more than a year, causing difficulties for the publisher. The authors express their sincere gratitude to Mr Ippei Ohta of Kodansha Scientific Ltd., Tokyo, Japan for his patience and encouragement throughout the translation of this work.

Koichi Nishikida
Etsuo Nishio
Robert W. Hannah

June, 1995

Part I
Fundamentals

1. Fourier Transform Infrared (FT-IR) System

1.1 Principle of the FT-IR Spectrometer

Infrared (IR) spectroscopy is concerned with the interaction between light and vibrational motion of the covalent chemical bonding of molecules and lattice vibrations of ionic crystals. The interaction energy corresponds to the so-called *infrared* light whose wavelength is *ca.* 0.7 – 1,000 micrometers or 10 to *ca.* 15,000 cm^{-1}. IR spectroscopy has been widely utilized by chemists because qualitative and quantitative information on the nature of chemical bonding and crystal lattices is deduced from transmitted, reflected, or scattered IR radiation.[1] The system used to measure those properties of the sample, *i.e.*, the infrared spectrometer, consists of an IR source, a device to decode the radiation into frequency and intensity information, a sample compartment, a detector, and a recorder device.

In a modern FT-IR system, IR radiation is introduced to a scanning interferometer and the output radiation intensity as a function of time is decoded into frequency and intensity information through a mathematical calculation known as Fourier transformation. The fundamental design of the interferometer is shown in Fig. 1.1.

Incoming IR radiation $I(\tilde{v})$ with wavenumber of $\tilde{v}$ cm^{-1} is divided by a beam splitter (BS) into two beams of equal intensity, the transmitted beam, $I_t(\tilde{v})$, and reflected beam, $I_r(\tilde{v})$. These two beams reflected respectively by two mirrors, fixed mirror M_1 and mobile mirror M_2, return to the BS again where the two beams recombine into one. Depending on the phase difference between the two returning beams, they are combined either constructively or destructively. When the pathlength difference between the two beams, x_1 and x_2, is equal to an integral number of wavelengths, the two beams with intensity $I/2$ have the same phase, and they combine constructively to give a beam of radiation with intensity I, as shown in Fig. 1.2(a). (However, since the BS reflects and transmits 50% of the radiation, the actual intensity reaching the detector is $I/2$, assuming 0 loss due to other causes such as reflection at the beam splitter surface.) However, if mirror M_2 moves in such a way that the difference between the respective optical path, often called the optical retardation, $\delta = x_1 - x_2$, becomes an odd multiple of half a wavelength, the two beams are destructively combined to leave zero intensity* as depicted in Fig. 1.2(b). This situation is first achieved when M_2 moves a quarter of the wavelength from the equidistant position (notice the path difference is twice that of the mirror displacement). If M_2 moves to a position corresponding to a distance equal to a half wavelength from the equidistant position, that is, δ is equal to the wavelength of the incident radiation, the two beams are again combined constructively to leave the intensity I, actually $I/2$, as noted previously. Therefore, as M_2 moves, the intensity of the monochromatic light of wavelength λ detected after the interferometer beam splitter is I when the path difference is $n\lambda$ ($n = 0, 1, 2, 3, \cdots$) and 0 when the path difference is $(n + 1/2)\lambda$. This procedure is described in Fig. 1.2.

The intensity of the output from the interferometer varies in a sinusoidal way with a frequency of

$$f = F\tilde{v} \text{ (Hz)} \tag{1.1}$$

---

* This does not mean the light itself disappears but the combined light transmitted through BS to return to the source has intensity of $I/2$. Therefore, the combined light going to detector and one returning to IR source are of 180° phase difference.

4

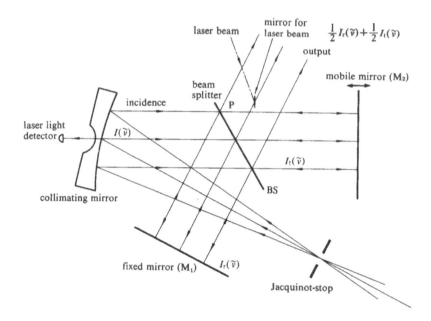

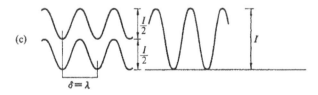

Fig. 1.1   Design of a traditional Michelson type interferometer.   (Courtesy of the Perkin-Elmer Corporation).

Constructive and Destructive Interactions at Beam Splitter

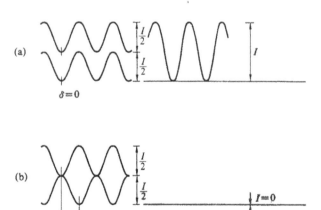

Fig. 1.2   Interference of two Infrared rays when the retardation is 0 (a), $2/\lambda$ (b), and λ (c).

where $\tilde{v}$ and F are wavenumber of the incoming radiation (cm^{-1}) and optical path difference velocity (OPD velocity; cm/sec), respectively.*

In order to understand the relationship between the interferogram and the input radiation, the following four interferograms will be considered.   Fig. 1.3(a) displays the interferogram (right) of monochromatic light (left) of frequency v_1 with no line width.   The frequency of the monochromatic light, v_1, and the frequency of the interferogram are related to each other, as described by Eq. 1.1.   If the frequency of the incoming light is high, the interferogram frequency is high and *vice versa*.   Fig. 1.3(b) is for the case of the same frequency of the light but with a finite line width.   It is clear from the comparison with case (a) that the interferogram intensity decays when the line has width.   If there are two monochromatic lines, both of which have no linewidth, the interferogram is a superposition of each interferogram as seen in Fig. 1.3(c).   For the same two lines having finite linewidths, the interferogram will be as shown in Fig. 1.3(d).   Although those four cases are not so difficult to comprehend, generally speaking it is not possible to visually estimate the spectrum from the interferogram.

To illustrate the more generalized case where the source emits radiation with a wide range of different wavelengths, Fig. 1.4 should be examined.   In Fig. 1.4, the interferograms of six different monochromatic lines of the same intensity, whose frequencies in wavenumber are $\tilde{v}_0$, $2\tilde{v}_0$, $3\tilde{v}_0$, $4\tilde{v}_0$, $5\tilde{v}_0$ and $6\tilde{v}_0$, (Fig. 1.4(a) ($\tilde{v}_0$) to Fig. 1.4(f) ($6\tilde{v}_0$)), and the sum of those six interferograms (Fig. 1.4(g)) is shown.   At $\delta = 0$, the intensity becomes very large, since all of the input frequencies return to the BS in phase and constructively combine.   This intense output is called the "centerburst" and the position at $\delta = 0$ is called "zero retardation."   The combined beam from the interferometer is introduced to the sample compartment where it experiences absorption, reflection, or scattering by the sample.   The radiation is detected by a detector and an optical signal is converted into an electric signal, which is further converted into digital numbers by an analogue-to-digital converter (ADC).   Thus, optical information is converted into electric digital information and is then Fourier-transformed into a spectrum.

Figure 1.5(a) is the interferogram of the single beam spectrum of the system itself.   This interferogram contains information pertaining to the IR radiation from the source, energy losses due to optical components, the profile of detector sensitivity, and CO_2 and H_2O vapor in the air, both of which absorb radiation.   IR radiation reaching the detector has a wide range of wavelengths with different intensities.   Thus, it is impossible to extract spectral data from this interferogram without the aid of a computer.

The weak signals which persist at large retardations as shown in Fig. 1.5(a) are due to spectral components with narrow bandwidths.   In the present case, most of the narrow lines are due to the absorption lines of CO_2 and H_2O shown in Fig. 1.5(b).   When the system is well purged, such that CO_2 and H_2O vapor are removed from the system atmosphere, the interferogram will decrease the intensity of the signals at large retardation and change to reflect those losses.

The system has two apertures, the beam splitter and second, either the Jacquinot stop (J-stop in Fig. 1.1), or the detector target.   The purpose of the second aperture is to limit the radiation to a more or less collimated beam at the beam splitter.   As shown in Fig. 1.6, the rays (2) which are α radians away from the collimated rays (1) have an optical path difference, $2l/\cos\alpha$, while that of the collimated radiation is $2l$, where l is the mirror displacement.   Since these two beams will interfere as explained with Fig. 1.4, the two beams will be combined destructively when the difference in the two retardations, $x = 2l\,(1/\cos\alpha - 1) \doteqdot l\alpha^2$, is a half wavelength of the

---

* In the case of a traditional Michelson interferometer, OPD velocity, F is equal to twice the mirror velocity, V. Thus, the equation $f = 2V\tilde{v}$ is given in many textbooks.   However, there are some different designs of interferometers in which the OPD velocity is not expressed by $2V$.   Thus, OPD velocity is more meaningful than mirror velocity in the discussion of many aspects of FT-IR spectroscopy.

6

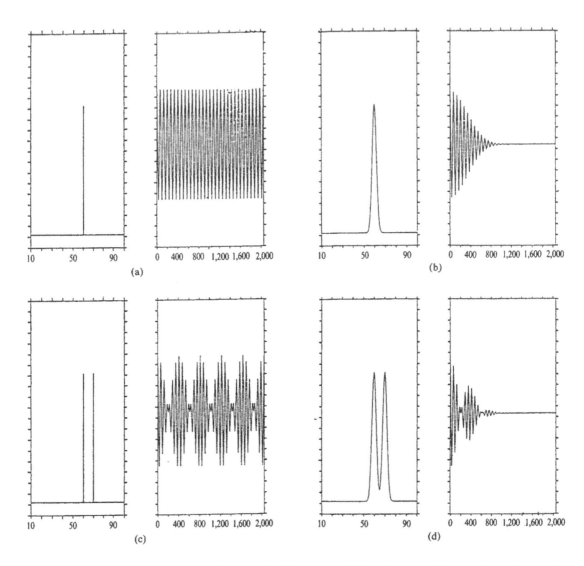

Fig. 1.3 Relationship between spectral lines and interferograms. (a) interferogram of a monochromatic ray of frequency ν_1 with no linewidth. (b) same as (a) but with finite linewidth. (c) interferogram of two monochromatic rays, ν_1 and ν_2 with the same intensity and no linewidth. (d) same as (c) but with the same finite linewidth to both rays.

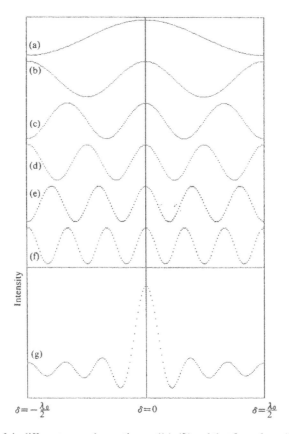

Fig. 1.4 Interferograms of six different monochromatic rays ((a)–(f)) and that from the radiation containing all the rays with the same intensity (e). Frequencies and intensities are described in the text.

radiation. This implies that the amplitude of the interferogram of the radiation with the wavelength of $2x$ decays to 0 at the retardation l. Thus, the resolution for the line with wavelength of $2x$ and shorter is not improved any more, even if the higher retardation is provided. Since the theoretical resolution, $\Delta\tilde{v}$, is given by[2)]

$$\Delta\tilde{v} = 1/(2l) \tag{1.2}$$

the maximum divergence, α_{max}, is given for the maximum frequency, $\tilde{v}_{max}$, and theoretical resolution as follows:

$$l\alpha_{max}^2 = 1/2\tilde{v}_{max} \tag{1.3}$$

Substituting for l from Eq. 1.2, $1/2\tilde{v}_{max} = \{\alpha^2/2\Delta\tilde{v}\}$, and

$$\alpha_{max} = \{\Delta\tilde{v}/\tilde{v}_{max}\}^{1/2} \tag{1.4}$$

Readers will understand that to move the mirror to the retardation necessary to achieve the theoretical resolution does not necessarily guarantee the resolution, unless the J-stop is closed to satisfy Eq. 1.4. Since the J-stop restricts the incoming radiation, a small J-stop for a high

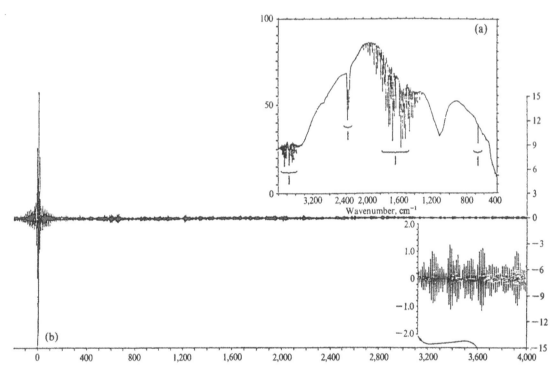

Fig. 1.5   Interferogram and a single beam spectrum taken from one of the commercially available FT-IR systems.

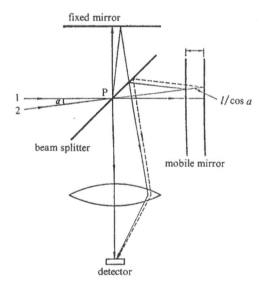

Fig. 1.6

resolution observation provides lower energy at the sample and detector, while the larger J-stop for low resolution observation provides higher energy.   The normal trade between signal-to-noise and resolution applies to the interferometer.

The laser radiation shown in Fig. 1.1 is fed into the interferometer and its interferogram is utilized to monitor the position of the moving mirror and determine the timing of data collection. Since the laser radiation is regarded as nearly monochromatic, *i.e.*, without linewidth, its interferogram is a well-defined sinusoidal wave which does not decay for large retardation. Thus, every singular point such as maxima, minima, and zero-crossing points of the interferogram can be utilized to determine the position of the moving mirror.

The interferogram of the laser radiation is also used to set the maximum wavenumber of the observation range of the system.   It is possible to define a sinusoidal wave which has amplitude I and frequency f or its overtones with frequency of nf, if the wave is measured at two different times.   Therefore, provided that the overtones are properly filtered out from the system, the shortest time interval of two sequential observations will determine the maximum frequency of the sinusoidal signal the system can define unequivocally.   This is given by the equation, $1/\Delta t_{min} = f_{max}$.   This situation also fulfills Nyquist's sampling theorem,[3] which states that in order to obtain complete statistical information regarding a measured signal, the signal must be measured at twice the highest frequency component.

Most of the commercially available FT-IR systems employ the so-called single beam optical path design.   Therefore, the normal operational mode is to record the system spectrum without sample followed by the sample spectrum, and to take a ratio of the two spectra.   This eliminates the background features of the instrumental system and leaves the spectrum due solely to the sample.

Although the above descriptions are the essential part of the FT-IR system, there are two other operations which are performed before the interferogram is actually decoded into a spectrum. The first is "phase correction" and the other is "apodization."

Phase correction will be discussed first.   The Fourier transform of a function $f(x)$ is given by[4]

$$F(\omega) = \int f(x)e^{-i\omega x}dx = \sum a(\tilde{v}_i)\cos\tilde{v}_i + i\sum b(\tilde{v}_i)\sin\tilde{v}_i$$

This yields only cosine terms when $f(x)$ is an even function like Fig. 1.4(g).   Thus, the Fourier transform is equivalent to the cosine transform and the spectrum is given by the cosine terms. On the other hand, if $f(x)$ is an odd function, the Fourier transform is equivalent to a sine transform and yields only the sine terms.   The interferogram of a real FT-IR system is neither an even nor odd function, and appreciable distortion occurs in the real interferogram, as shown in Fig. 1.5(a).   Fourier transform of the interferogram will therefore yield both cosine and sine terms.   When there are non-zero sine terms, the spectrum given solely with a cosine transform only is not a true spectrum and is distorted.   It is neccessary to account for the existence of the sine terms.   The first method is a "phase correction" method.   A small number of points in the interferogram around the center burst is first Fourier transformed and the "phase angle ϕ," the ratio of sine and cosine terms at each data point, is calculated.   Using these phase angles, the

hat range is corrected, so that the sine terms disappear from the Fourier
rrected interferogram.[5]   It is sufficient to correct only the central part of the
e points removed even a few points from the centerburst are of weak intensity
ute to the distortion.

od is to perform the Fourier transformation of the asymmetric interferogram
bsolute value or power value defined by $A(v) = (a(v)^2 + b(v)^2)^{1/2}$.   With this
um is given by $A(v)$ and is called the "power spectrum" or "magnitude
method leads to a positive shift in the zero transmission position, since

squaring the noise which is both positive and negative around zero transmission leads to positive numbers only.

Now, let us discuss the second problem, the apodization.   As shown in Fig. 1.4, an interferogram of any monochromatic line, whose linewidth is infinitisimally small, is a sinusoidal wave continuing forever.   To be consistent with this fact, as expressed with Eqs. 1.2 and 1.3, the Fourier transform requests the integration limits go from a retardation δ of $-\infty$ to $+\infty$. However, since it is physically impossibly to scan the interferometer from $-\infty$ retardation to $+\infty$ retardation, we have to restrict the traveling distance of the moving mirror.   As illustrated later in Fig. 1.6, the integration of Eq. 1.3 over the restricted range from $-L$ to L instead of $-\infty$ to $+\infty$ gives rise to a line with a finite linewidth accompanied by a series of sidebands in the wings of the absorption band at lower and higher frequencies.   If we multiply the interferogram with a function which decreases in its magnitude as the retardation increases in such a way that the magnitude of the interferogram becomes zero at $\pm L$, the transform of a modified interferogram will show smaller side-lobes when the integration omits the range over $\pm L$.   While the side-lobes decrease, the line-width becomes wider, and band intensity weaker.   The process of multiplying the interferogram by a function whose purpose is to reduce or remove these side-lobes is called *apodization*.

Any reader who is interested in more details concerning the basic theory of Fourier Transform Infrared Spectrometer systems should read the textbook cited in reference.[6]   In this book the authors are concerned mainly with how to set the instrument parameters, including the sampling accessory, in order to optimize the system for a given analysis.

1.2   Scan Conditions

Before one scans the FT-IR instrument, the scan conditions must be established.   The parameters which should be set by the operator are: (i) resolution, (ii) combination of detector, source, and beam splitter to match the frequency range, (iii) OPD (optical path difference) velocity, (iv) apodization function, and (v) choice of phase correction.   The combination of parameters should be chosen to match the analytical or physical problem being solved.   In addition, the choice of a suitable accessory and reference material for background observation is another important factor.

1.2.1   Resolution
The theoretical resolution of an FT-IR system is given by

$$\Delta \tilde{v} = 1/2L \; (\text{cm}^{-1}) \tag{1.4}$$

where $2L$ is the optical retardation.   Therefore, if one changes the resolution from 4 to 2 cm^{-1}, the moving mirror must travel twice as far and the time needed for a scan will be two times longer. In addition, higher resolution requires the use of a smaller size aperture, J-stop, as discussed previously, so that significant amounts of the IR energy are blocked by the J-stop and a poorer signal-to-noise ratio of the spectrum results.   In order to recover a higher signal-to-noise ratio, it is necessary to average a larger number of scans, or to scan more slowly.   Therefore, it may be wasteful of time to employ an unnecessarily high resolution.   Since the mean halfwidth for condensed phase samples is 6 cm^{-1}, the present authors recommed 2 cm^{-1} resolution for liquid samples and also for special cases of solid samples such as crystalline materials and most of the low molecular weight solid samples.   Apart from some of the crystalline polymers such as polyethylene, which may require 4 cm^{-1} or even 2 cm^{-1} resolution for detailed characterization,

4 or 8 cm^{-1} resolution should be sufficient to analyze most polymers and plastics.   An 8 cm^{-1} resolution is commonly used for low energy applications such as GC-IR and microscope analysis.

The most controversial case is gas analysis.   Since the IR spectra of light diatomic and triatomic molecules such as HF, HCl, NO, CO, H_2O, and CO_2 are split into well-developed fine structures due to rotational interaction, one may conclude it is essential to have resolution as high as 0.2 or even 0.1 cm^{-1}.   Indeed, it is needed to have such a high resolution capability for the physicochemical studies of gaseous samples for precise determination of the molecular structure or Coriolis constants.[7]   However, it should be emphasised that from the authors' experience, moderate resolution such as 1 or 2 cm^{-1} is best suited for most cases of *quantitative* analysis of gaseous samples, in order to provide a reasonable compromise between the signal-to-noise and linearity.   In other words, to use resolutions better than 1 cm^{-1} may result in a serious degradation in quantitation because of lower signal-to-noise.

1.2.2   Detector
A.   Mid IR Detector

The DTGS (Deuterated Tri-Glycine Sulfate) pyroelectric detector is a standard detector of the FT-IR spectrometer.   Since the response time is of the order of a few μsec,[8] the OPD velocity of most of the commercially available FT-IR spectrometers using this detector are about 0.2 – 0.4 cm/sec in order to avoid detector saturation effects.   However, surveying the FT-IR spectrometers, the OPD velocities provided for the use with DTGS detector range from 0.05, the slowest, to 1.2 cm/sec, the fastest.   The DTGS detector has the characteristic that the output voltage increases when the OPD velocity is decreased.[8]   Roughly speaking, the output becomes two times larger when the OPD velocity is halved, keeping the noise level the same.   Therefore, the default OPD velocity, more or less a fast one, is set for high transmission samples such as polymer films or liquid samples in a liquid cell.   However, when the amount of IR radiation becomes small due to the nature of lower transmission accessories such as diffuse reflection, ATR, or an IR microscope, or due to high absorption by black samples such as used engine oil or tires, it is recommended that slow OPD velocities be used.   For example, two scans with an OPD velocity will improve S/N value $\sqrt{2}x$ better than one scan, while one scan with an OPD velocity which is 1/2 of the previous speed will give a S/N value $2x$ better.   Therefore, one scan with this slower OPD speed will give $\sqrt{2}$ times better signal-to-noise than the two scans using the initial (faster) scanning speed.

The MCT (Mercury-Cadmium-Telluride) detector[9] works only when the detector element is cooled to low temperature, for example to liquid nitrogen temperature.   Thus, this detector is mounted in a metal or glass Dewar vessel to be cooled with liquid nitrogen at 77 K.   In contrast to the DTGS detector, the response time is short and as a result the OPD velocity must be faster, once again to avoid saturation.   In addition, this detector inherently has higher signal-to-noise. Since the MCT detector is saturated easily by incident radiation, it is best used for low energy level rvations such as GC-IR and the IR microscope.   However, when one needs an MCT tor to detect a very weak signal from a high transmission sample, one must pay attention on vel of the incident radiation intensity in order to avoid detector saturation.   The following vill be instructive to check if the detector system is saturated.   Measure the transmission rum of a polystyrene film and attenuate the IR radiation using a neutral density filter such reen.   Repeat the observation with a new background.   If the relative intensities of the rption bands change after attenuation, the MCT detector is indeed saturated. pical OPD velocities for the MCT detector range from 0.5 cm/sec to 5.0 cm/sec.   Since the MCT detector shows poor signal-to-noise and saturation with slow OPD velocities, the lower OPD velocities provided for the DTGS detector in general, should not be selected.

There are three basic types of MCT detectors depending on the low frequency cut-off.   The narrow range MCT detector whose cut-off frequency is *ca.* 750 cm^{-1} provides the highest signal-to-noise.   The wide range MCT detector has a lower cut-off frequency of *ca.* 400 cm^{-1} covers an optical range as wide as DTGS detector.   The wide range detector has a lower signal-to-noise than the narrow range detector by almost one order of magnitude, but has a higher signal-to-noise than the TGS detector.   The medium range (600 cm^{-1} cut-off) MCT detector sits in between the narrow and wide range detectors in terms of signal-to-noise and range.   The reader must also be aware that MCT detector suppliers provide different size detector elements.   The most commonly used are 1×1 mm, but 0.5×0.5 mm elements and sometimes smaller depending on the application are available.   The user should consult with the operator's manual supplied from the FT-IR manufacturers for the choice of target size.   Large elements such as 1×1 mm element are used for transmission, specular reflection, ATR, and/or diffuse reflection as well as emission spectroscopy, because the full aperture of the IR beam is utilized.   On the other hand small elements such as 0.25×0.25 mm and 0.1×0.1 mm elements are used for GC-IR and IR microscope measurement, because in those accessories the IR beam is focused to the small size and the small element is needed to match to the small image size of those focused beams.   If the small image is focused on the large element, the signal created by the image is diluted by the noise from the element area where no IR beam arrives.   It is possible, therefore, to use even a 0.05×0.05 mm element for IR microscopy measurements.   This sort of micro element will give very high optical signals for small area observation in the microscope, for example 10×10 μm samples.   However, the reader should be aware that optical adjustment of the IR microscope with very small size detectors is extremely difficult.

B.   Far IR (FIR) Detector

A DTGS detector with a special window material which is transparent in the FIR region, 700 – 10 cm^{-1}, has sensitivity in the FIR region.   Polyethylene is a typical window material for a standard FIR detector.   Typical OPD velocities are from 0.1 to 0.2 cm/sec, which is about the same speed as the mid-IR region.

A Si-bolometer[10] is a high sensitivity detector for the FIR region.   This detector element is operative at liquid helium temperature, 4.2 K.   A high vacuum system is required with a proper vacuum gauge in order to evacuate the liquid helium container of the Si-bolometer.   Furthermore, liquid nitrogen is required to insulate the helium container of the Si-bolometer and the vacuum system also requires the use of liquid nitrogen.   Because of the rather large size of the liquid helium container, the distance from the window to the detector element is far enough to make an optical adjustment tedious.   In this regard, the use of the parabolic light guide inside the bolometer is recommended.   The light guide introduces the IR beam down to the element once the IR beam hits the detector window, even if the direction of the beam is away from the element.   The use of the light guide will be essential when an IR microscope is connected to FIR-FTIR system.

The Si-bolometer is easily saturated by radiation.   Since the human body emits FIR radiation, the detector can easily be saturated by stray radiation from the body.   Proper shielding should be provided.   OPD velocity of the Si-bolometer is faster than that of the DTGS detector.   Typical OPD velocities are from *ca.* 0.5 – 2 cm/sec.

C.   Near IR (NIR) Detector

The DTGS detector has sensitivity in this region as well and is commonly used.   If high sensitivity is required, then cooled detectors[11] such as InAs, InSb, and InGaAs detectors are available.   These detectors are most sensitive at low temperatures such as liquid nitrogen temperature, and are mounted in Dewer vessels like the MCT detector.   All detectors are easily saturated with low to moderate radiation energy amounts similar to the MCT detector and Si-bolometer.   In fact the suitable level for those detectors is less than 4% of the incident radiation for InAs and InSb detectors and less than 1% for InGaAs detector and these detectors should be used with a combination of low throughput accessories such as diffuse reflection or optical fibers. For example, for open beam application such as transmission spectra of solid films, a TGS detector or cooled detector with the radiation attenuator may give almost the same signal-to-noise for a given time.[11]   In this case there is no significant advantage to using cooled detectors. Although InAs and InSb detectors have similar characteristics, InSb has slightly higher cut-off frequency or narrower range with better signal-to-noise in 10,000 to 7000 cm^{-1} range compared with InAs detector.   Because of the smaller dynamic range of the InGaAs detector, it is usually used as a detector for the FT-Raman system.   Appropriate OPD velocity for those detectors is within 0.2 – 2 cm/sec.   Although high speed application like GC-FTIR application is not common for the NIR region, higher OPD speed benefits users for higher signal-to-noise for a given time in the case of low energy applications.

1.2.3   Apodization

As discussed in section 1.1, the apodization function is introduced to reduce the amplitude of a sideband, which appears in a spectral band when an interferogram for the radiation does not decay within a given retardation.   Therefore, certain apodization is needed when the linewidth is smaller than instrumental resolution, which is given by $\Delta \tilde{v} = 1/2L$ (L = retardation).   For example, suppose a case in which a gas phase line of, say, 0.10 cm^{-1} linewidth was measured with 0.40 cm^{-1} resolution.   Let us call this situation "under-resolved."   The opposite to this situation is called "well-resolved".   This will be exemplified by a condensed phase line with 6 cm^{-1} linewidth measured with 2 cm^{-1} instrumental resolution.   Since the interferogram for a radiation decays within a retardation in this case, side-lobes will not appear in a spectrum. Therefore, there will be no need of apodization.   Indeed, apodization will affect the spectrum of the "well-resolved" case with loss of spectral information.

The effect of apodization on the under-resolved water vapor line is illustrated in Fig. 1.7. These are of water vapor calculated from the same interferogram taken with a theoretical resolution of 0.50 cm^{-1} but using several different apodization functions.   The sidebands which appear in the case of non-apodization (Fig. 1.7(a); sometimes called "boxcar apodization") marked by arrows disappear in the case of triangular apodization and the peak height becomes 1/2 of the unapodized spectrum (Fig. 1.7(f)).   A set of apodization functions are proposed in order to remove the sidebands with a minimum of line broadening.   The reader will find it quite clear that the sidebands are reduced or disappear at the expense of resolution.   The effect of the apodization is seen not only on gaseous samples but also on liquid and solid samples.   Fig. 1.8(a), 1.8(b) and 1.8(c) are taken with boxcar and weak and strong Norton-Beer apodization functions, respectively.   This example illustrates how the selection of the apodization function affects the spectral feature of the polyethylene CH_2 bending vibrations observed as a doublet at 1470 and 1460 cm^{-1} measured even with 8 cm^{-1} resolution!

Hannah[12] examined the effect of apodization on signal and noise and line shape, using indene as an example.   Fig. 1.9 shows the effect of apodization for (a) "under-resolved" and (b) "well-resolved" cases.   In both cases, the linewidth increases when the apodization function changes from "none-apodized" to "strong Norton-Beer" function, depressing the peak height as expected.   The reader will notice the side-lobes in the case of boxcar apodization for the "under-resolved"/ case.   Since noise may be considered as an aggregation of many narrow lines, apodization will also suppress the amplitude of noise.   It should be noted, however, that the noise and signal are affected differently by apodization.   As listed in Fig. 1.7, signal-to-noise increased as a stronger apodization function is used, while the linewidth increased.   Apodization causes trade-off

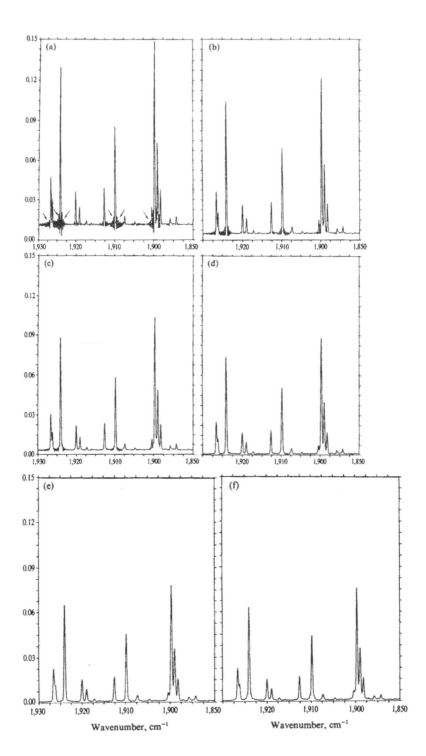

Fig. 1.7 FT-IR spectra of water vapor calculated using different apodization functions. (a) non-apodized (boxcar) (b) raised cosine, (c) Norton-Beer weak, (d) Norton-Beer medium, (e) Norton-Beer strong, and (f) triangular.

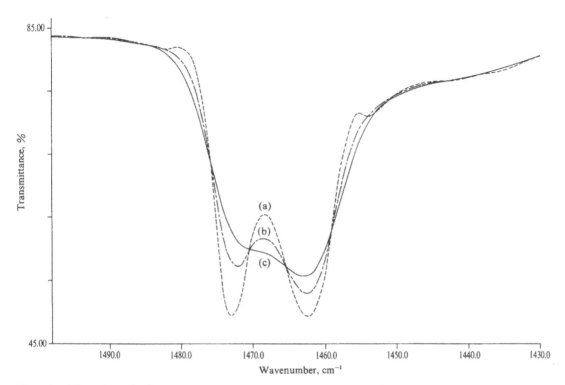

Fig. 1.8 Effect of apodization on absorption bands (1,470 and 1,460 cm⁻¹) of poly(ethylene) at 8 cm⁻¹ resolution. (a) boxcar, (b) Norton-Beer weak, and (c) Norton-Beer strong.

between resolution (or linewidth) and signal-to-noise. As expected the peak area stayed more or less constant with different apodization functions.

The effect of apodization is not only observed on the spectral features such as linewidth and sidebands. Spectroscopically, the more important effect of apodization[13] is on the linearity at high absorbance. For instance, for an absorption band whose full width at half height is 0.1 of the peak height, boxcar apodization will give a linear relationship between true and apparent absorbance values when the true absorbance is small. As the true absorbance value approaches 0.5, the apparent absorbance starts leveling off. However, as the true value continues to become larger, the apparent value suddenly rises to infinity. In the case of triangular apodization, after the linear relationship between the real and apparent absorptions for weak absorption, the apparent value levels off to a value as the true value grows. Thus, the spectrometer will fail to provide meaningful value for intense bands. Griffith[6a] recommends the use of the Norton-Beer weak function[14] for high resolution or when good quantitative analysis is required. Since the apodization function is found to affect the quantitization of sample, one must calibrate the system using suitable standards.

Thus, it would be an improper conclusion if the reader thought that some of the FT-IR spectrometers gave better or worse spectra when some differences in spectral features such as peak intensity and resolution were observed from the same sample taken on different manufacture's FT-IR spectrometers, when the difference is in reality due to apodization difference.

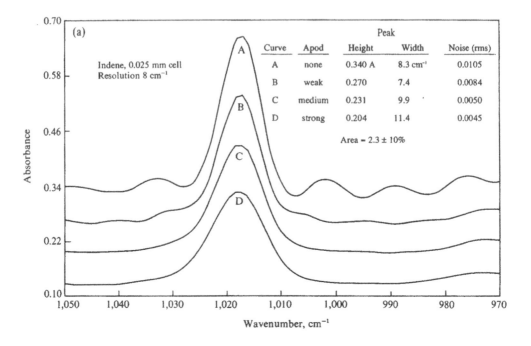

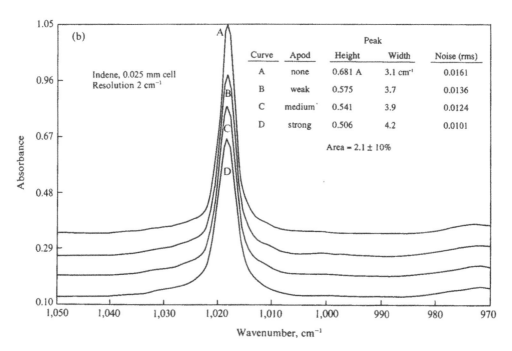

Fig. 1.9   Effect of apodization on the indene spectrum of under-resolved (a) and well-resolved (b) cases.

1.2.4   Phase Correction and Magnitude Spectra

As explained in Section 1.1, the phase correction is neccessary to obtain an undistorted spectrum from a realistic FT-IR system.   Also, it was shown that another method, "magnitude spectrum" or "power spectrum" method, is available to obtain an undistorted spectrum without phase correction.   The latter method has the advantage that the spectrum can be calculated when the signal is quite weak to perform phase correction or when an extremely large phase shift is observed as in the case of the photoacoustic (PA) detector.   However, the disadvantage of this method is that the ordinate value is always shifted to a higher value at zero transmission, as explained above.   This trend becomes conspicuous when the weak signal or dark sample accompanied by high noise is measured.   Therefore, the magnitude method is recommended only when the PA detector is used, or in those other cases of low signal levels.

NOTES AND REFERENCES

1. (a) W.J. Moore, *Physical Chemistry*, Prentice Hall, New York (1955); (b) L. Pauling and E.B. Wilson, *Introduction to Quantum Mechanics*, Chapt. 10, McGraw Hill, New York (1935); (c) K. Nakamoto, *Infrared Spectra of Inorganic and Coordination Compounds*, Wiley Interscience, New York (1963); (d) I.J. Bellamy, *Advances in Infrared Group Frequencies*, Methuen (1968); (e) A. L. Smith, *Applied Infrared Spectroscopy*, Wiley, New York (1979).
2. P.R. Griffith and J.A. de Haseth, *Fourier Transform Infrared Spectrometry*, pp. 9 – 15, Wiley Interscience, New York (1986).
3. M. Woodward, *Probability and Information Theory*, Pergamon Press, New York (1955).
4. R.N. Bracewell, *The Fourier Transform and its Applications*, McGraw Hill, New York (1986).
5. L. Marz, *Transformations in Optics*, Wiley, New York (1965); L. Marz, *Infrared Phys.*, 7, 17 (1967).
6. (a) ref. (2); (b) R.J. Bell, *Introductory Fourier Transform Spectroscopy*, Wiley, New York (1977); (c) *Fourier Transform Infrared Spectroscopy*, Vols. 1-4, (ed. J.R. Ferraro, L.J. Basile) Academic Press, New York (1978).
7. G. Herzberg, *Infrared and Raman Spectra of Polyatomic Molecules*, Van Nostrand Co., New York (1945).
8. Technical Data Sheets (TGS), Mullard Co., Southampton, U.K. (1900).
9. For instance, read "EC & G Judson Infrared Detector," Brochure, pp. 20 – 28, EG & G Judson, Montgomeryville, PA. (1900).
10. For Si-bolometer, refer to "Product Brochure," Infrared Laboratories Inc., Tucson, Arizona.
11. For Near-IR detector (InSb, InAs), refer to "EG & G Judson Infrared Detectors" Brochure, pp. 2 – 19, EG & G Judson, Montgomeryville, PA. (1900).
12. R.W. Hannah, Unpublished data.
13. R.J. Anderson and P.R. Griffiths, *Anal. Chem.*, 47, 2339 (1975); also refer to Chapter 10 of ref. (2).
14. R.H. Norton and R. Beer, *J. Opt. Soc. Am.*, 66, 259 (1976).

2. Features and Operation Techniques of the Infrared Accessories

2.1 Infrared Microscope

2.1.1 Principle

The term "infrared microscope" may sound like a microscope to magnify the image of the sample using infrared light consistent with the usage applied to an electron microscope, which utilizes the electron beam to magnify the image of the sample. However, it is actually a high magnification beam condenser system which introduces the focused IR radiation to a small part of the transmitting or reflecting sample in order to obtain its IR spectrum. Since it is necessary to observe the sample visually in order to select that part of the sample whose IR spectrum is to be obtained, an optical microscope is coupled to this system. For most of the commercially available IR microscopes, IR and visible radiation share part of the microscope optical system to condense and transfer the IR and visible radiation to a small area of the sample. Since glass or quartz does not transmit mid- and far-IR radiation, lens optics made of glass or quartz are not usable. In addition, some materials such as KBr, CsI, and KRS-5 which transmit both IR and visible light are not suited for lenses, because of their properties. For example, they are brittle, hygroscopic, or colored. Furthermore, since the refractive index is changing, not all wavelengths are focused property. Thus, instead of a lens system, a reflecting magnifying mirror system such as a Cassegrain configuration (detail shown in Fig. 2.1) is used for the IR microscope.

The optical design of one of the commercially available IR microscopes is shown in Fig. 2.1. For the observation of an IR transmission spectrum, the infrared radiation from the FT-IR optical unit is focused by a Cassegrain (C_1) on a plane (F_1) where the sample is located. The radiation transmitted through the sample is focused on another plane (F_2) by another Cassegrain (C_2), so that the real image of the sample is formed at F_2. The image is further transferred to the detector element, where the sample response is detected. In order to investigate a designated area of the sample, a variable aperture is placed at F_2 to restrict the radiation arriving at the detector to that from the unmasked position of the sample. In this low energy situation the MCT detector offers better performance than the DTGS detector.

Although almost all of the commercially available infrared microscopes employ a more-or-less similar design to that described for transmitting samples, optical designs for a reflecting sample are rather different from supplier to supplier. In the case shown in Fig. 2.1, the incident radiation is diverted by mirror (M_1) and the radiation is focused by Cassegrain (C_2) on sample (F_1) and the reflected radiation is collected by the same Cassegrain (C_2) and focused in the F_2 plane. The radiation which passes through aperture is thus focused on the detector as described in the case of transmitting samples. Examples of other optical designs for reflecting samples are shown in Fig. 2.2. In the case of the arrangement shown in Fig. 2.2(a), half the incident radiation, passing through the aperture, is focused on the reflecting sample utilizing one half of the Cassegrain (C) and the reflected radiation is collected by the other side of the Cassegrain (C).

A special modification to the Cassegrain is done in the case of Fig. 2.2(b). In this design, a small mirror is placed in the central space of the Cassegrain, so that the small mirror collimates the incident radiation and half of the Cassegrain is utilized to collect the reflected radiation. Fig. 2.2(c) is a type different from the previous design, in which the incident radiation is collimated by a rather large mirror and the reflected radiation is collected from one side of the Cassegrain. A

20

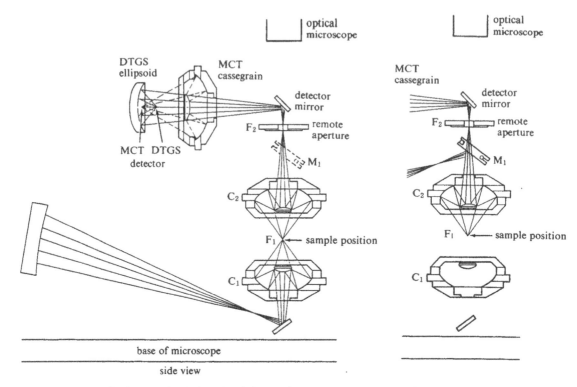

Fig. 2.1 Optical design of an infrared microscope. (Courtesy of Perkin-Elmer Corp.)

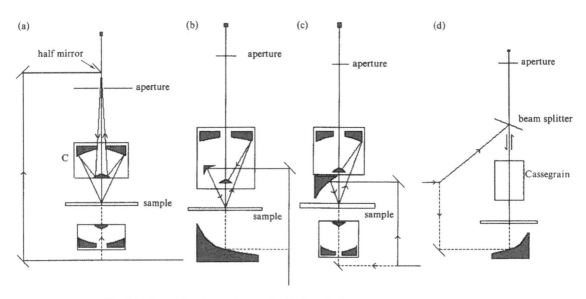

Fig. 2.2 Examples of reflection mode of infrared microscopes.
———: Path for reflection mode, ----: Path for transmission mode, ■: Detector

beam splitter is used to accommodate reflection capability in the case of Fig. 2.2(d).

In addition the traditional modes, transmission and reflection, of observation, devices for microscopic ATR[1] (see Section 2.2 for the principle of ATR) and grazing angle reflection[2] (see Section 2.5 for the principle) studies are available. Readers will find some applications of these techniques in Part II of this monograph.

When visual selection of the observation area is attempted, visible light replaces the IR radiation at the sample. The visible light transmitted or reflected from the sample is introduced to a binocular-type viewer instead of being led to the detector. Therefore, the IR radiation and the visible light must be tuned well to pass the same optical path at the sample. A pair of optical polarizers may be placed before and after the sample in the visible light path. Those facilities are provided for users to obtain a *visual image* of the sample under polarized light. User should consult the textbook[3] on optical microscopes for the applications of polarizers. For IR dichroic microscopic studies of transmitting and reflecting samples, the *IR polarizer* (not the same polarizer for the optical polarizers!) should be placed in the position the supplier suggests.

2.1.2   Experimental Procedure

In the procedure to observe the IR spectra of the sample, the most important consideration is how to match the optical and instrumental conditions of background and sample observations. The size, shape and position of the aperture and all of the instrumental settings except the number of scans must be precisely maintained during background and sample observations, because any incompensation may give rise to artifacts, as will be shown later. The sample area of observation must be located at the center of the field of vision.

The next step is to determine the order of background and sample observations. Sample spectra can be obtained by either of the following two methods. The first method is more or less a standard FT-IR observation method in which the single beam background spectrum is measured first and stored in the memory of the system. Then the sample is measured to provide its IR spectrum, because the system will automatically divide the single beam spectrum by the stored background spectrum.

The second method requires the operator's involvement. The single beam spectra of both the background and sample are determined independently, with the sample observation probably done first. The operator performs the division of the sample single beam spectrum with background single beam spectrum on the system computer later on. The second method may be more convenient when positioning of the sample requires tedious work.

First, switch the system into visual microscope mode. Observe the sample and bring the target area into the center of the field of vision, with the aid of the cross-hair mark carved in the eye-piece lens. Unlike the optical microscope, focusing is conducted by moving the sample stage up and down. Determine the area of observation by setting the aperture to highlight the area. Then determine the position for background observation by moving only the sample stage horizontally, so that the open space where there is no sample is brought into the aperture (Do not touch the aperture). Once decisions on the position and the order of sample and background observations are made, bring the sample or reference area into the aperture for the first observation and switch the system into IR observation mode.

Since the refractive index of the sample is different from that of air and visual focusing can be done for the upper or bottom surface of the sample, further adjustment must be carried out for the best IR energy throughput after visual focusing. In order to do this, change the vertical position of the condenser Cassegrain (C_1 in Fig. 2.1) slightly in the case of transmission or vertical position of the sample stage in the case of reflection measurement, so that the energy throughput will be optimized. Then finish the first observation and switch the system into optical

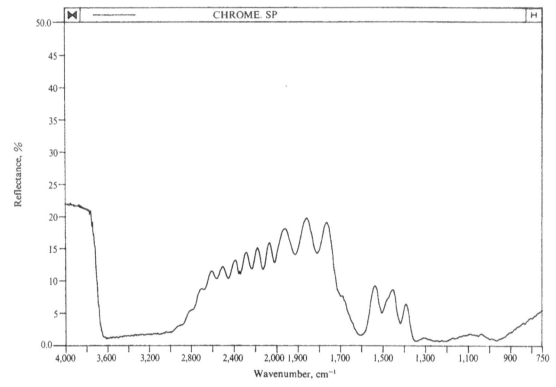

Fig. 2.3   An IR spectrum of a chrome-black coating material.

microscope mode.   Bring the other position into aperture and adjust the energy for the subsequent observation.

If there is no position for background material in the case of the reflecting sample, the aluminum or gold mirror provided by the system manufacturer can be used as reference material. In this case one must change the reference and sample for each observation.

Since the metallic surface in the optical path reflects IR radiation, the surface is chemically treated so as to tint black.   However, no matter how deep black the parts are tinted, a part of the radiation is reflected by the surface and this will leave the spectral feature of the black layer when it arrives at the detector.   One example of incompensation most probably due to misadjustment of the optical setting is shown in Fig. 2.3; this is due to the thin layer of chrome black coating commonly used to tint metal black.

2.1.3   Features of IR Microspectroscopy

As explained in Section 2.1.1, the IR microscope utilizes Cassegrain optics and aperture(s). As shown in Fig. 2.1, the incident angle to the sample is not normal.   Moreover, it has a range, the minimum and maximum of which are available from the supplier or can be calculated from the minimum and maximum numerical apertures (N.A.) of the Cassegrain used.   When a narrow slit is placed on the pass of the light, it causes diffraction of the light and as a result the beam spreads out to the physical shadow.   In other words, contribution of spurious light outside the aperture should be taken into account extracting information from the ordinate values.

Those phenomena cause certain optical effects which may cause the transmittance and reflectance values to deviate from the true values.   Thus, some spectroscopic data can be erroneous without the proper correction. Another source of error is non-uniformity of the samples. In this section we will discuss the special features inherent to IR microscopy.

A. Effect of wide incident angle distribution of cassegrain

Fraser[4] in 1953 showed that the dichroic ratio (see Chapter 3, Section 3.1 for the fundamental concepts regarding dichroic ratio measurements) of a uniaxially drawn sample determined from the IR spectra taken on an IR microscope must be corrected using the minimum and the maximum N.A. as follows

$$D = R_{//} / R_{\perp} = \log[T_{//} + M(T_{\perp} - T_{//})]/[\log T_{\perp}] \tag{2.1}$$

$$M = 1 - 3/2(\cos \theta_1 + \cos \theta_2)/\cos^2 \theta_1 + \cos^2 \theta_2 + \cos \theta_1 \cos \theta_2$$

$$\sin \theta_1 = (N.A.)_1/n, \quad \sin \theta_2 = (N.A.)_2/n$$

where n, $(N.A.)_1$, and $(N.A.)_2$ are the refractive index of the sample, and the minimum and the maximum numerical apertures, respectively. For the Cassegrain used in Fig. 2.2(a), θ_1 and θ_2 are 15.68 and 35.5 for a 15× magnification Cassegrain and 15.7 and 40.5 for a 32× magnification Cassegrain[5] respectively. For example, the dichroic ratio of a peak whose transmittance is 75%T and 70%T with parallel and perpendicular polarization is 0.72 with correction or 0.81 without correction by 32× Cassegrain. However, the error becomes very large when the dichroic ratio deviates substantially from 1. From instance, the dichroic ratio of a peak whose parallel absorption is 50%T and perpendicular absorption is 90%T using the 32× Cassegrain is 11.2 with correction but 6.58 without correction. Thus, experimental dichroic data obtained with an IR microscope could be substantially in error when the analyte peak shows significant dichroism.

B. Effect of apertures on spacial resolution and stray light

Using the model of Fraunhofer diffraction,[6] we will discuss the effect of the slit in IR microspectroscopy. Let us consider the optical arrangement shown in Fig. 2.4(a), where the parallel light after mirror system (M_1) passes through the aperture and is focused on the screen (C). In a real IR microscope, the aperture is located where the real image of the sample instead of the source is formed, and the image will be focused on the detector element instead of the screen. Using the wave equation,[7] the amplitude dy_0 of the central light at P and that (dy_s) of the light at P passing the point "s" of the aperture S are given by

$$dy_0 = (Ids/x)\sin(\omega t - kx)$$

$$dy_s = (Ids/x)\sin(\omega t - kx - ks \sin \theta) \tag{2.2}$$

respectively. Adding all of the light beam passing $+b/2$ through $-b/2$, it can be shown that the total amplitude of y is given by

$$y = (Ib/x) \cdot \sin(0.5kb \sin \theta) \cdot \sin(\omega t - kx)/(0.5kb \sin \theta) \tag{2.3}$$

This equation is rewritten as

$$A = A_0(\sin \beta / \beta)\sin(\omega t - kx) \tag{2.4}$$

where $\beta = 1/2kb \sin \theta = (\pi b/\lambda)\sin \theta$ and $A_0 = Ib/x$. Thus, intensity of the light is given by the square of A, or

$$A^2 = A_0^2 \sin^2 \beta / \beta^2 \tag{2.5}$$

where a time average of $\sin(\omega t - kx)$ is given by $<\sin^2(\omega t - kx)> = 1/2$. Fig. 2.4(b) shows the intensity, A^2, of the single slit diffraction pattern. Point P_1 in Fig. 2.4(a) is a position where the

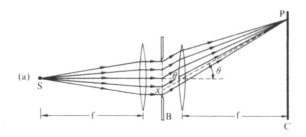

(a)

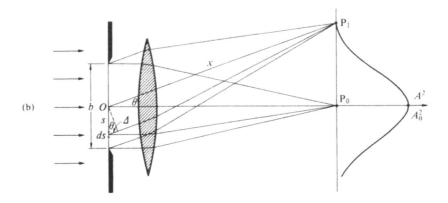

(b)

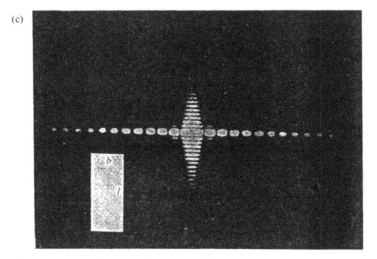

(c)

Fig. 2.4 Effect of a narrow slit in an optical path.
(a) parallel rays passing through an aperture focused on a screen
(b) light intensity of the single slit diffraction
(c) a photographic image of a diffraction caused by a square slit.

rays at $+s$ and $-s$ to the center will destructively interfere with each other to leave 0 intensity. For instance, it is evident from Fig. 2.4(b) that the ray passing the lower edge has a longer path than the central ray by nearly $\lambda/2$ and the ray from the upper edge has a path length shorter than the central one by nearly $\lambda/2$. Thus, intensity becomes 0 when $\theta = (\lambda/b)$. This indicates that the frequency of the progression is high and decay of the progression is rapid when the slit width is wide, while the frequency of the progression is low and the decay is slow when the slit width is narrow. The effect of the slit width is well demonstrated in Fig. 2.4(c), in which two different series of the $\sin \beta/\beta$ progressions were observed.

When the slit width is much larger than wavelength, for example, $b = 10\lambda$, the image of the slit is well defined as shown in Fig. 2.5(a). When the slit width is five times larger than wavelength (Fig. 2.5(b)) the image of the slit broadens. When the width of the slit becomes comparable to the wavelength, the diffraction becomes significant, and it is not possible to define the image of the slit, as shown in Fig. 2.5(c).

The same progression will be generated by rays adjacent to the edge of the slit which degrades the spatial resolution. When the contribution of the light intensity from progression becomes equal to that from the image, the spatial resolution is lost completely. Thus, closing the aperture to less than $10\,\mu m$, which is equivalent to the wavelength at $1,000\ cm^{-1}$, will not improve the discrimination of the target area. The addition of another slit will reduce[8] the spurious radiation nearly 50%. However, additional aperture further restricts the throughput of the IR radiation, making observation more difficult.

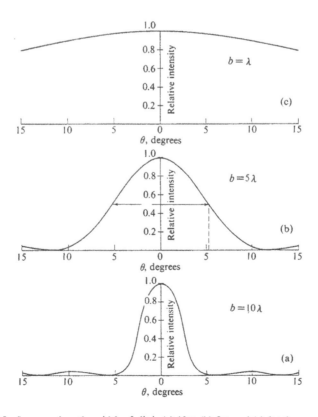

Fig. 2.5 Images when the width of slit is (a) 10×, (b) 5×, and (c) 1× the wavelength.

C.   Irregular sample shapes

Many of the subjects for IR spectroscopic studies have irregular or non-uniform shapes, since these subjects are fibers, irregular parts of uniform films, or foreign matter and deposits on regularly shaped products such as Si wafers, hard disk media, and plastic products.

Consider an irregularly shaped chip of a solid as an example.   This material is approximated as wedge[9] when the thickness of each slice of the sample is rearranged from the smallest to the largest thickness, as shown in Fig. 2.6.   In addition, stray light due to diffraction will not be eliminated easily, because the small aperture will reduce the sensitivity quite significantly, and, in addition, the irregular solid may have a hole.   Therefore, the intensity of an absorption band will be the average for the optical path length through the sample, *i.e.*,

$$A(\tilde{\nu}) \backsim \int d(x) \mathrm{d}x \tag{2.6}$$

It was concluded that only weak absorption bands[10] can be utilized for quantitative analysis.

Thus, IR microscope should be used primarily for qualitative analysis.   When IR microscopy is to be used for quantitative analysis, careful calibration must be performed.

In addition to this, as explained in Section 3(a), only the weak absorption bands can be utilized for quantitative purposes such a dichroic ratio determinations.   Absorption bands showing average dichroism may not be utilized.   It was also shown in Section 3(b) that an aperture to outline the area of measurement will introduce stray light, becoming another cause of error in quantization.

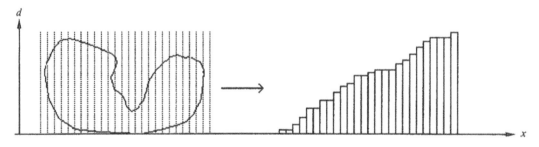

Fig. 2.6   Shape of a non-uniform sample (a) and a quasi-wedge obtained by rearranging slices (b).
d: thickness, x: displacement

2.2   Attenuated Total Reflection (ATR), or Multiple Internal Reflection (MIR) Accessory[11]

2.2.1   Principle

In the ATR method the sample is placed in contact with the surface of an optical element (called the "Internal Reflection Element," or IRE) having a high refractive index.   Light is trapped within the IRE by the phenomenon called total internal reflection and produces *evanescent wave* (E-wave; *vide infra*) at the boundary between the prism and sample; this wave penetrates a very small distance into the sample.   The resultant transmission of the radiation which passes through the IRE changes due to the sample *absorption*.   By this means, the ATR technique gives rise to absorption-like spectra which are, in fact, somewhat different in some spectral aspects from

those of an absorption spectrum obtained by transmission.   Since the details of the ATR method
have been described in detail by Harrick and others,[12] only the minimum of the theory and
practice of the ATR technique is given here in order to allow users to utilize the technique
efficiently.

First, let us discuss the reflection of light at the boundary between media 1 and 2 whose
refractive indices are n_1 and n_2 ($n_1 > n_2$), respectively.   When the incident radiation in medium
1 reaches the boundary with medium 2 at an angle greater than the so-called critical angle, defined
as $\theta_c = \sin^{-1} (n_2/n_1)$, total reflection occurs.[12]   Both parallel and perpendicular polarized light
will be totally reflected at any incident angle equal to and greater than the critical angle, θ_c, as
shown in Fig. 2.7(a).   The term internal reflection means that the light in the higher refractive
index medium is reflected at the boundary with lower refractive index medium.   It is necessary
to clearly distinguish between *total internal reflection* as described above, and *external reflection*
(Fig. 2.7(b)) in which the radiation in the lower refractive index medium 2 reaches the boundry
to medium 1 having higher refractive index.

Although it is said that 100% of the incident radiation is reflected in the case of total internal
reflection, microscopically speaking, the incident radiation penetrates into medium 2 by a
fraction of a wavelength.   The radiation thus penetrating into medium 2 is called the *evanescent
wave*, as mentioned above.   Readers should remember that an evanescent wave decays
exponentially as Fig. 2.8 shows even if the medium 2 is a nonabsorbing medium.   This property
of the evanescent wave is in contrast with the IR radiation of the IR source in the IR spectrometer,
which does not decay when it travels through nonabsorbing and nonscattering media.

If we define the penetration depth, d_p, of the evanescent wave as the distance at which the
intensity of the evanescent wave becomes $1/e$, then d_p is given by Eq. 2.7, provided that the sample
(medium 2) does not have absorption at the wavelength of the incident radiation.

$$d_p = (\lambda/2\pi n_1)(\sin^2 \theta - (n_2/n_1)^2)^{-1/2} \tag{2.7}$$

where $\lambda, \theta, n_1, n_2$ are the wavelength of the incident radiation, the incident angle, and the refractive

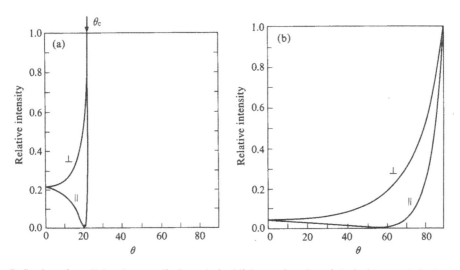

Fig. 2.7   Reflection of parallel and perpendicular polarized light as a function of the incident angle in the case of (a)
internal reflection (n_1=4.0 (Ge) and n_2=1.5 (general organic compound)) and (b) external reflection (n_1=1.0
(air) and n_2=1.5).

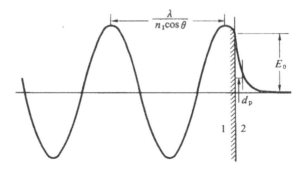

Fig. 2.8   Exponentially decaying electric field of an evanescent wave at the boundary to the totally reflecting material. In the dense medium (1), the electric field which has a sinusoidal dependence was projected to the normal line.

indices of the IRE and sample, respectively.   According to Eq. 2.7, the penetration depth becomes smaller as the incident angle becomes larger.   Also, Eq. 2.7 indicates that the penetration depth becomes smaller as the refractive index of the IRE becomes larger.

The absorbance in the ATR spectrum of a sample having thickness, d, much greater than d_p, for 1 reflection is given in absorbance unit as

$$A = \log e(n_2/n_1)(E_0^2/\cos\theta)\int\exp(-2z/d_p)dz \tag{2.8}$$

or

$$A = \log e(n_2/n_1)(E_0^2/\cos\theta)(d_p/2)\alpha \quad \text{(for } d \longrightarrow \infty) \tag{2.9}$$

where E_0 is an electric field defined by n_1, n_2 and θ, and α is the extinction coefficient of the sample for unit thickness.   Eq. 2.9 together with Eq. 2.7 states that the intensity of the ATR spectrum will be larger when the incident angle is smaller and also when the difference between n_2 and n_1 is smaller, but $n_1 > n_2$.

2.2.2   ATR Accessories

Commercially available ATR accessories fall into three categories, vertical ATR, horizontal ATR, and cylindrical ATR accessories.   In the case of vertical ATR accessories, the sample and the IRE are vertically placed and have the highest versatility in experimental conditions such as incident angles, dichroism experiments, and absorption intensities.   Shapes of IRE are trapezoidal, parallelogram, and hemi-cylindrical types.   Horizontal ATR accessories are designed for the easiest use and the quickest sampling.   Cylindrical ATR accessories are primarily for the observation of liquid samples while the former two types are primarily for solid samples.   The most popular ATR accessory today is the horizontal ATR (HATR) accessory.

A.   Vertical type ATR accessories

There are two different optical designs in this category, Wilks type and Harrick type.   Optical design of the Wilks type accessory is shown in Fig. 2.9(a).   As indicated by the arrow, one can change the incident angle by changing the position of IRE, which is a trapezoidal shape.   There are two kinds of Wilks type accessories, one which allows continuously variable positions of the IRE and the other which restricts the IRE positions for specific incident angles such as 30°, 45°, and 60°.   In both cases the accessories are designed to allow multiple reflection in order to enhance absorption intensity.   Commercially available IREs are made of KRS-5, ZnSe, Si, AgCl, AgBr, and Ge.   The face angle of the IRE is 45° or 60° in the cases of KRS-5 and ZnSe. Si and Ge IRE are supplied with 30°, 45° or 60° face angles.   In order to select a specific incident

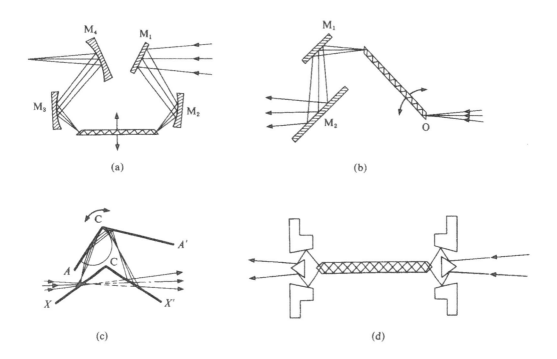

Fig. 2.9   Optical diagrams of Wilks type (a), Harrick type (b), (c) hemispherical IRE type, and (d) cylindrical IRE type ATR accessories.

angle, the IRE with the face angle equal to the incident angle must be placed at the position marked as that incident angle.   For example, if the reader wishes to select the 45° angle of incidence, an IRE with a face angle of 45° must be placed at the position marked 45°.   When an IRE with the length, l, thickness, d, and the face angle, α, is placed on the position marked for an incident angle, β, of the Wilks type accessory, the effective incident angle, θ, and the number of reflections, N, are given by

$$\theta = \alpha + \sin^{-1}((\beta - \alpha)/n_1) \tag{2.10}$$

$$N = (1/d)\cot\theta \tag{2.11}$$

In Table 2.1, the effective incident angle, θ, the number of reflection in parentheses, and the penetration depth, d_p, are listed for the combination of face angle and positioning.

Si should be a quite useful material for the IRE, since the refractive index of Si, 3.4, lies between those of KRS-5 and Ge.   Unfortunately, Si has strong absorptions below 1,500 cm^{-1} and this may set the low frequency limit to $ca.$ 1,500 cm^{-1} depending on the quality of Si used as the IRE.

A design of the Harrick type vertical ATR accessory is shown in Fig. 2.9(b), in which the incident angle is selected continuously by rotation of the parallelogram-shaped IRE.   In this method, the number written on the IRE position selector has a different meaning depending on the face angle and refractive index of the IRE.   The reader must consult with the operator's manual to set the incident angle.

Hemi-cylindrical IRE accessory, whose optical diagram is shown in Fig. 2.9(c), has the

TABLE 2.1

material of IRE and face angle		set position	30°	45°	60°
KRS-5	45°	effective incident angle (number of reflections) d_p	39° (–) do not use	45° (25.0) 0.20 λ	51.3° (20.0) 0.14 λ
KRS-5	60°	effective incident angle (number of reflections) d_p	47.8° (22.7) 0.17 λ	53.7° (18.4) 0.13 λ	60° (14.4) 0.11 λ
Si	30°	effective incident angle (number of reflections) d_p	30° (43.3) 0.177 λ	34.2° (36.8) 0.125 λ	38.2° (31.8) 0.102 λ
Si	45°	effective incident angle (number of reflections) d_p	40.8° (29.0) 0.092 λ	45° (25) 0.081 λ	49.2° (21.6) 0.073 λ
Ge	30°	effective incident angle (number of reflections) d_p	30° (43.3) 0.12 λ	33.7° (37.5) 0.097 λ	37.2° (33.0) 0.084 λ
Ge	45°	effective incident angle (number of reflections) d_p	41.3° (28.5) 0.073 λ	45° (25) 0.066 λ	48.7° (22.0) 0.061 λ

d_p : penetration depth   ▢ standard setting

advantages of semi-micro sampling and the continuously variable incident angle setting, and is the best suited for depth-profiling descrebed in section 4.3 of this chapter.   However, this accessory provides only one reflection, giving rather weak absorption intensities compared with those accessories offering multiple reflection capabilities.   Although this type of ATR accessory was widely used in the beginning of ATR application, only Harrick Scientific supplies this type of ATR accessory today.

  B.   Horizontal type ATR accessories

  In the case of horizontal ATR accessories, the IRE is placed horizontally, and the sample is placed horizontally on it.   This allows easy sampling for a variety of samples including liquids, powders, and pellets.   Most of the commercially available accessories employ thick IRE to allow a forced contact between IRE and solid samples such as powders, granules, and pellets.   On the other hand, sampling is allowed only on one surface of the IRE.   The number of reflections inside the IRE is smaller than for vertical ATR accessories, because the horizontal ATR accessories employ thick IREs.   This causes weaker absorption intensities than vertical ATR accessories.   In addition, unlike vertical ATR accessories, it is not able to cover a wide range of incident angles.*

  C.   Cylindrical IRE type ATR and related accessories

  Although the ATR accessory utilizing a cylindrical IRE was initially introduced by Spectra-Tech Corp. as "CIRCLE CELL™", Axiom Analytical Corp. also supplies a similar accessory, "TUNNEL CELL™".   As shown in Fig. 2.9(d) for the case of "CIRCLE CELL™", this type of the accessory utilizes a cylindrical IRE for liquid samples.   Instead of utilizing a cylindrical IRE, SPECAC offers a similar ATR accessory for liquid samples using a square IRE.   For use as the detector for liquid chromatography, all suppliers provide heated pressure-proof micro-flow cell type versions.   Further development of the ATR accessory for liquid samples has prompted some accessories for monitoring chemical reactions *in-situ* and/or quality monitoring of large-scale samples.   Those are, for example, the "DIPPER CELL™" from Axoim, the ATR needle-

---

* Pike Technologies supplies a variable angle HATR accessory.

probe from Spectra-Tech, and the "Process Flow ATR system" from SPECAC. However, it should be added that Harrick Scientific developed the first ATR accessory for monitoring liquid products or liquid stream in 1971, utilizing a trapezoidal IRE.

2.2.3   Experimental Procedure

An example of how to use an ATR accessory, using a three-position trapezoidal type one as an example, will be given. Since total reflection is achieved only when the incident angle is larger than the critical angle, *i.e.*, $\theta > \theta_c$, θ_c must be known when the combination of IRE, KRS-5 ($n_1 = 2.4$), ZnSe ($n_1 = 2.4$), Si ($n_1 = 3.4$), or Ge ($n_1 = 4.0$), and the sample is selected. Since the refractive index of the majority of organic compounds such as resins and solvents lies between 1.4 – 1.6, θ_c will be $\approx 38°$ when KRS-5 or ZnSe prism is selected as the IRE. θ_c will be $\approx 23°$ for the combination of an organic compound and a Ge IRE. Therefore, as a minimum requirement, KRS-5 or ZnSe prism with a face angle of 45° should not be placed at the 30° position.

It is recommended that the single beam spectrum of the system equipped with ATR accessory be measured as a background spectrum, since the frequency dependence of the optical throughput of the accessory, including transmission characteristics of IRE, will be involved in the background spectrum. For the best optical throughput, two mirrors (M_2 and M_3) must be carefully adjusted, so that the incident radiation falls on the side face of the IRE most efficiently and the radiation transmitted through IRE must be most efficiently collimated on the detector. Utilize the laser light of the interferometer as a guide. Repeat the procedure until the largest throughput of the energy is achieved.

The sample must be cut such that the pieces cover the entire surface of the IRE as shown in Fig. 2.10(a), or cover the part of IRE as shown in Fig. 2.10(b) when enough sample is not available. It is important to cover the short axis of the IRE entirely to avoid the contribution of radiation which does not interact with the sample. The single beam spectrum of the sample must be measured without changing any of the experimental settings, such as the incidence angle. By division of the sample spectrum by the background spectrum, which is done automatically for most FT-IR systems, the ATR spectrum of the sample is obtained.

As the ATR technique is concerned with the optical phenomena at the boundary between the sample and the IRE, intimate contact between sample and IRE is essential to obtain quality spectra. Therefore, liquid samples are the easiest to observe the ATR spectra. A flexible, continuous solid, such as certain polymer films is also a favorable sample for the ATR method. In order to achieve good contact between sample and IRE, however, elastic materials such as soft rubber pads may be placed between the sample and back plates, so that the uniform pressure on the sample assures uniform, intimate contact. On the other hand, non-flexible solid, especially

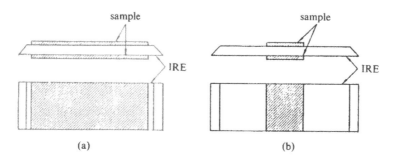

Fig. 2.10   Sampling methods for a general sample (a) and a small sample (b).

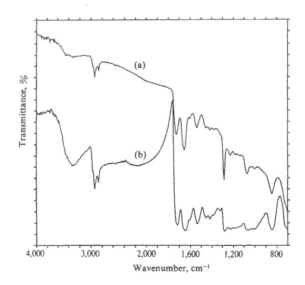

Fig. 2.11   ATR spectra of video tape when measured with a correct setting (Ge IRE and 45° angle of incidence) (a) and
an incorrect setting (KRS-5 and 45° angle of incidence) (b).   (Ordinate expanded)

powder of such non-flexible sample is not a favorable sample, because air gap will cause the rapid
decay of the evanescent wave since d_p for air ($n_2 = 1.0$) is far smaller than for sample ($n_2 = 1.5$).   It
is possible to observe quite good ATR spectra when the air gap is filled with liquid such as mineral
oil, say Nujol.   The HATR can also be used to observe powder samples when a device is used to
press the powder to IRE.

It should be noted, however, that the observed spectra could show artifacts resulting in an
erroneous observation when a high refractive index compound such as carbon black is mixed in
the sample.   Fig. 2.11 illustrates the correct and incorrect choices of the IRE and angle setting
for the ATR observation of an inorganic filled magnetic layer of a video tape.   While spectrum
(a) is taken with a 45° angle of incidence Ge IRE and all of the peaks show the feature of
absorption bands, spectrum (b) taken with a 45° angle of incidence KRS-5 IRE shows band
distortions at ~1730, ~1280, ~1080, and ~850 cm^{-1}.   Thus, spectrum (b) *cannot* be used for
qualitative and quantitative analysis.   The cause of these distortions is due to the large refractive
index of the magnetic layer of the video tape.   ATR observations of carbon-containing "dark
samples" such as floppy disk media, black rubber O-rings, and tires should be performed with a
Ge IRE and probably 45° or higher angle of incidence.

2.2.4   Features of ATR Spectra
A.   General features
Readers must be aware that the following three characteristic features are always associated
with ATR spectra.

(1) Since the penetration depth is proportional to the wavelength as stated by Eq. 2.7,
absorption intensity is also proportional to the wavelength.   In other words, peaks at low
frequencies are relatively stronger than peaks at high frequencies.

(2) Since the refractive index of the sample abruptly changes before and after the peak position
(this phenomenon is called "anomalous dispersion of refractive index".   See Fig. 2.23 for

example), some of the ATR peak positions would be shifted *lower* frequency from those of a transmission spectrum as one would expect from Eq. 2.9, and the relative peak intensity would be different from that of a transmission spectrum.   In an extreme case, artifacts such as non-existing peaks may appear in the ATR spectrum.   Examples will be presented shortly.

(3) Since the penetration depths of parallel and perpendicular light are different (refer to Section 3.1D), peak intensities will be different for ATR spectra observed with the parallel and the perpendicular polarized light, even if the sample is not oriented.

Items (1) and (2) will be discussed in more detail, for these two are the most important features of the ATR technique, while item (3) must be remembered in order to perform correct dichroic ratio experiments by the ATR method.   It must also be remembered that item (3) is a cause of the non-linear calibration behavior between concentration and ATR intensity.[13]

Consider Fig. 2.12.   Comparing the peak intensities of a triplet, a singlet and a doublet at respectively 2,900, 1740, and 600 cm^{-1} in an ATR spectrum of plastisized poly(vinylchloride) (Fig. 2.12(a); (KRS-5, 45° angle of incidence)) with those observed in a transmission spectrum of the same sample (Fig. 2.12(c)), the reader will find it obvious that the peak intensities in the low frequency region are relatively stronger than those in the high frequency region.   Assuming Eq. 2.7 to give the relative penetration depth of the ATR spectrum, one can make corrections on the spectrum shown in Fig. 2.12(a) by dividing the ordinate values by the wavelength to give a spectrum whose penetration depth is uniform over the entire abscissa, as shown in Fig. 2.12(b). Although Fig. 2.12(b) and 2.12(c) coincide fairly well, rigorously speaking, peak ratios of the 2900

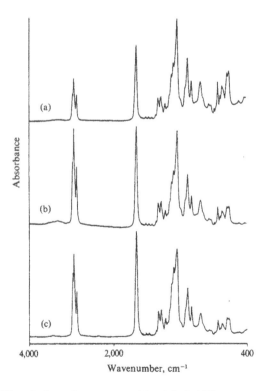

Fig. 2.12   Comparison of ATR and absorption spectra.   (a) original ATR spectrum of poly(vinyl chloride) with plasticizer, (b) corrected for the wavelength dependence of penetration depth, and (c) absorption spectrum of the same vinyl chloride sample.

and 1740 cm^{-1} bands for those two spectra are not exactly the same.   This suggests that Eq. 2.7, which describes the ATR spectra with one parameter, such as penetration depth calculated with n_1 and n_2, both of which are deemed as constant, cannot precisely describe ATR spectra.   At the same time, fairly good coincidence of Figs. 2.12(b) and 2.12(c) implies that we can explain most of the features of ATR spectra in terms of Eq. 2.7.   Thus, Eq. 2.7 is the most important equation in the ATR technique.

On the other hand, Eq. 2.9 describes the ATR spectra not only with penetration depth but also in terms of $(n_2/n_1)\alpha$.   Since n_1 can be regarded as a constant for a given IRE in a non-absorbing region, Eq. 2.9 is equivalent to Eq. 2.7, if we assume n_2 is a constant.   However, because of the anomalous dispersion, the sample refractive index, n_2, exhibits maximum and minimum values at lower and higher frequencies of the peak position, respectively.   Therefore, the peak position of the ATR spectrum given by $(n_2/n_1)\alpha$ will be shifted towards a lower frequency (or red-shift) from that of a transmission spectrum given solely by α, and this shift becomes non-negligible in the case of strong absorption peaks.   The shift and the appearance of artifacts in the vicinity of strong bands in ATR spectra are shown in Fig. 2.13.   Among the three ATR spectra of ester bands of a polycarbonate, Fig. 2.13(c), which is taken with Ge IRE at a 45° angle of incidence shows the best coincidence with the transmission spectrum.   As the incident angle approaches the critical angle estimated from the refractive index of ZnSe, 2.37, and the average value of the polycarbonate refractive index, $ca.$ 1.5, for example Figs. 2.13(b) and 2.13(a), the discrepancies in frequencies and intensities between ATR spectra and transmission spectram become more conspicuous.   For example, in Fig. 2.13(a) observed by ZnSe IRE at a 45° angle of incidence the ATR spectrum showed splitting of the 1,164 cm^{-1} band into two peaks.   This splitting cannot be attributed to a chemical cause such as special orientation of the polycarbonate at the surface and must be attributed to an artifact,[14] because Figs. 2.13(b) and 2.13(c) did not show the splitting when the incident angle was set farther from the critical angle and therefore a shallower

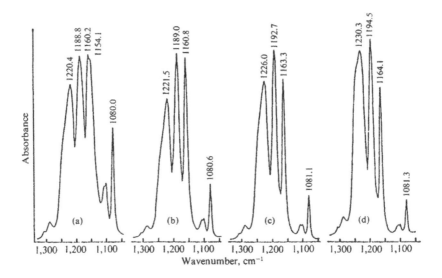

Fig. 2.13   ATR spectra of a polycarbonate film under different experimental conditions.   (a) KRS-5 IRE with 45° angle of incidence.   (b) KRS-5 IRE with 60° angle of incidence.   (c) Ge IRE with 45° angle of incidence. (d) transmission spectrum of the same polycarbonate film.

penetration of the surface of the sample than for the case of Fig. 2.13(a) is obtained.   Thus, for the precise determination of peak position and intensity, the use of Ge with a 45° angle of incidence instead of KRS-5 at 45° angle of incidence is recommended, although the Ge spectrum is significantly weaker in peak intensities compared with those with the most common case of KRS-5 at 45° angle of incidence.

One must know the value of n_2 and apply it to Eq. 2.9 to simulate ATR spectrum,[15] in order to seek a better coincidence with the observed absorption spectrum of Fig. 2.12(c) than the corrected one of Fig. 2.12(a).   One can calculate n_2 from the absorption or reflection spectrum, using Kramers-Kronig integration.   (Refer Fig. 2.23(c) for the refractive index, n, calculated by Kramers-Kronig analysis.)

B.   Surface analysis

As Eqs. 2.7 and 2.9 indicate, one can calculate and set the penetration depth, by selecting the IRE material and the incident angle.   This allows us to calculate the thickness of the surface layer in the case of well-defined bilayer samples.   Assume the peak (absorbance) intensities of individual ATR spectra of the surface material and the supporting substrates as $A_s(\infty)$ and $A_b(\infty)$, respectively, provided that the thickness of both surface material and substrate are much larger than the wavelength of interest.   Also, assume that the ATR peak intensities of the surface material and substrate of the double layer with a surface thickness of d are $A_s(d)$ and $A_b(d)$, respectively.   Iwamoto and his group showed that the thickness of the surface layer of a well-defined bilayer sample was satisfactorily calculated from the following two equations[13]

$$\ln[(A_s(\infty) - A_s(d))/A_s(\infty)] = -2d/d_p \tag{2.12a}$$

$$\ln[A_b(d)/A_b(\infty)] = -2d/d_p \tag{2.12b}$$

C.   Depth profiling

As mentioned above, one can set the depth of the observation by changing IRE material and incident angle.   This creates an opportunity to "determine the chemical composition as a function of depth," *i.e., depth profiling*.   When the chemical composition, $c(z)$, of the sample is a function of the depth z, the absorbance of the ATR spectrum as given by Eq. 2.8 is modified as follows:

$$A(2/d_p) = \log e(n_2/n_1)(E_0^2/\cos\theta)\int c(z)\exp(-2z/d_p)\mathrm{d}z \tag{2.13}$$

This equation implies that the *Laplace transform* of $c(z)$ is $A(2/d_p)$.

Thus, Eq. 2.13 indicates to us the possibility that $c(z)$ can be determined by the *inverse Laplace transform* of a series of $A(2/d_p)$, which are obtained by a series of ATR experiments with different d_p achieved by using different IRE's and different incident angles.

This calculation was done in 1975 by Hirschfeld[16].   Since this method is believed to be tedious, a simulation method was employed,[17] in which an analytical function was assumed for $c(z)$, so that the Laplace transform of $c(z)$, *i.e.*, the integration of Eq. 2.13, can be found easily in a mathematics textbook.   Those who are interested in the ATR depth profiling should read those original papers as well as recent ones.[18]

D.   ATR difference spectroscopy

The algorithm for spectral subtraction between two ATR spectra depends on the morphology of the sample surface.   When the ATR spectrum of a uniform mixture of two components, A + B, is subtracted with that of a pure component, B, both of which are taken with the same IRE and incident angle, the spectral subtraction is straightforward.   Since both samples are homogeneous, the wavelength dependence of the penetration depth is common to both spectra.   Thus, a single subtraction factor which is constant over the entire frequency region will eliminate the spectrum due to component B from the ATR spectrum of A + B.[*1]

On the other hand, if the sample has a well-defined bilayer structure,[*2] subtraction of an ATR spectrum of pure surface material, A_s, or pure substrate, A_b, from the ATR spectrum of a bilayer, A, taken with the same IRE and the same incidence angle for the purpose of getting substrate or surface material will be frustrated, because the penetration depth depends on the frequency, while the thickness of the surface is constant.

If the reader tries to obtain the surface spectrum by subtracting the ATR spectrum of the substrate (thickness $\gg d_p$) from the bilayer ATR spectrum (the same IRE and incident angle for both spectra), using the FT-IR manufacturers' spectral subtraction software, the higher frequency portion of the substrate spectrum will be over-subtracted and the lower frequency portion of the substrate spectrum will be under-subtracted. If the spectral subtraction is attempted to obtain the substrate spectrum by subtracting the ATR spectrum of the thick (thickness $\gg d_p$) surface material from the bilayer spectrum, the higher frequency portion of the surface spectrum is under-subtracted and the lower frequency portion of the substrate spectrum is over-subtracted.   Fig. 2.14 exemplifies these two cases.   In the attempt to obtain the spectrum of the surface material (polystyrene) of polyurethane coated with 0.31 μm thick polystyrene, the 1,731, 1,702, and 1,532 cm^{-1} peaks due to polyurethane are over-subtracted in the difference spectrum Fig. 2.14(b).   On the other hand, the 1,029 and 907 cm^{-1} peaks of polyurethane are under-subtracted.   Fig. 2.14(c) is the opposite case to Fig. 2.14(b).   In the attempt to obtain the substrate (polyurethane) spectrum, 1,493 and 1,452 cm^{-1} peaks due to polystyrene are under-subtracted and 750 and 698 cm^{-1} peaks are over-subtracted in the difference spectrum.   The vertical bars in Fig. 2.14(b) and (c) are the position where the scaling factors for the subtractions were calculated.   Thus, it is clear that a single subtraction factor which is a constant value over the entire spectral range is not a valid factor over the full range.

It is necessary, therefore, to consider the effect of the wavelength dependence of the penetration depth on the ATR difference spectroscopy.   Since the E-wave decays from 1 to $\exp(-2d/d_p)$ as it travels distance d, the absorbance due to the substrate in the bilayer sample is given by

$$A_b(d) = A_b(0) \exp(-2d/d_p) \tag{2.14}$$

where $A_b(0)$ is the absorbance due to the substrate when there is no surface material.   By spectral subtraction, the spectrum due to surface, A_s, is given by

$$A_s = A - f_b A_b \tag{2.15}$$

where A and f_b are an absorbance of the bilayer sample and subtraction factor, respectively. Since A_b should be equal to $A_b(d)$, f_b is given by $f_b = (A_b(0)/A_b)\exp(-2d/d_p)$.   Rewriting d_p as

$$d_p = d_p{}^0\lambda = d_p{}^0/\tilde{v}$$

Thus, Eq. 2.15 gives

$$A_s = A - c_2\exp(-c_1\tilde{v})A_b \tag{2.16}$$

Using the same procedure, the spectral subtraction to obtain the substrate spectrum will be given by

---

[*1] However, some of the absorption lines of both components may not be well eliminated, leaving derivative type absorption bands after subtraction.   This is due to the shifts in the peak positions in the case of the mixture due to a refractive index effect or molecular interaction between those components such as hydrogen-bonding or a charge-transfer interaction.

[*2] The bilayer structure we are concerned with is limited to the case where the thickness d of the surface layer is within the following condition; $0 < d \leq 3d_p$, because the E-wave will decay to 0.25% at the boundary when the surface is thicker than $3d_p$, making the contribution of the substrate spectrum negligible.

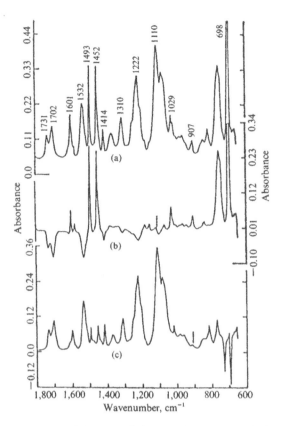

Fig. 2.14 A difference spectrum between ATR spectra of bilayer and substrate to calculate the spectrum of the surface material (a) and a difference spectrum between ATR spectra of bilayer and surface material to calculate the spectrum of the substrate.

$$A_b = A - f_s A_s = A - c_2[1 - \exp(c_1 \tilde{v})]A_s \tag{2.17}$$

In order to derive this equation, it should be understood that the absorbance due to the surface material whose thickness is d in the bilayer is described by

$$A_s(d) = A_s(\infty)[1 - \exp(-2d/d_p)] \tag{2.18}$$

Eq. 2.17 and Eq. 2.18 were derived[12b)] under the assumption that the decay of the E-wave due to the absorption by the surface material and substrate is negligible.   Since this is not true in regions of strong absorption, the difference spectroscopy for bilayer compounds is extremely difficult to be obtained by ATR technique.

E.   ATR spectra of very thin films

So far, the features of the ATR technique were developed under the assumption that the sample thickness or the surface layer thickness, d, is of an order of penetration depth of the evanescent wave, d_p, or thicker.   However, when the thickness of the sample or the thickness of the surface layer of a bilayer substance is very thin, $i.e.$, $d \ll d_p$, it is necessary to treat the ATR phenomenon differently.   When the sample or the surface layer of the bilayer compound is thin, the evanescent wave interacts more with the air or the substrate.   Thus, the total reflection phenomenon includes three substances, the IRE (n_1), the surface layer (n_2), and the substrate or air (n_3). Harrick[19)] gave effective thickness of the surface layer for parallel and perpendicular polarization as follows

$$d_{/\!/} = \frac{4n_{21}d\cos\theta[(1+n_{32}{}^4)\sin^2\theta - n_{31}{}^2]}{(1-n_{31}{}^2)[(1+n_{31}{}^2)\sin^2\theta - n_{31}{}^2]} \qquad (2.19a)$$

and

$$d_{\perp} = \frac{4n_{21}d\cos\theta}{(1-n_{31}{}^2)} \qquad (2.19b)$$

where d is a thickness of the thin film and $n_{ij} = n_i/n_j$.   It is clear that medium 1 and medium 3 play more role in the total reflection than medium 1 and medium 2.

2.3   Diffuse Reflection Accessory

2.3.1   Principle

The diffuse reflection technique became quite popular after the introduction of the FT-IR technique, because IR spectra of powder samples are easily measured by simply putting the neat powder or powder diluted in KBr or KCl powder in the sample cup.   However, the technique itself was used in the 1930s.[21]   The phenomenon of "diffuse reflection" is that in which the incident radiation is scattered in all directions by repeated reflections and refractions at the surface of particles, as illustrated in Fig. 2.15.   The diffusely reflected radiation contains a specular reflection component as marked by an arrow, a multiply refracted component, multiply reflected component, and other components which undergo a series of reflections and refractions. Although the specular reflection component comes from the very first layer of the sample, the path length of the rest of the individual IR rays vary because each ray of incident radiation is scattered differently by particles in the sample powder.   It has been reported that the radiation

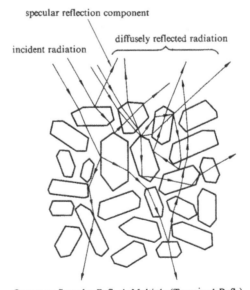

Output = Specular Refl. + Multiple (Trans and Refl.)

Fig. 2.15   Schematic optical diagrams of diffuse reflection.

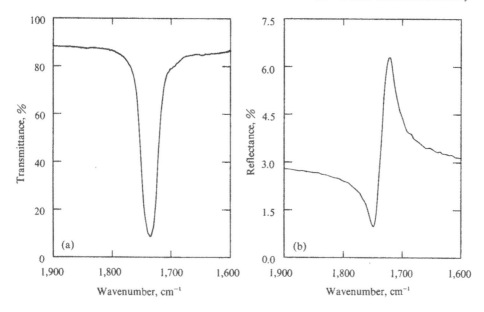

Fig. 2.16   Relationship between absorption and reflection spectra of an oscillator at 1736 cm⁻¹.

reaches *ca.* 3 mm deep from the surface.[22]   Therefore, it should be remembered that the diffuse reflection spectrum always contains both absorption and reflection spectral components, which are shown in Fig. 2.16(a), the absorption spectrum, and (b), the reflection spectrum of the oscillator with frequency $\tilde{v}$.   In addition, as mentioned above, the path length of the sample is not uniquely given for a diffuse reflection observation.

2.3.2   Experimental Procedure
Fill the sampling cup with KBr or KCl powder and place it in the diffuse reflectance accessory. A single beam spectrum of the system equipped with the accessory must be measured and used as the background spectrum.   Then dilute the sample powder by carefully mixing with KBr or KCl powder to a certain concentration (*vide infra*) and put in the sample cup.   The single beam spectrum of the system with the sample is taken and divided by the background spectrum to give the sample diffuse reflection spectrum.   In these two procedures, the height of the sample cup must be carefully adjusted so that the energy throughput is optimized.   The same KBr or KCl powder must be used for background and sample measurements.

For the far-infrared (FIR) region of observation, poly(ethylene) powder can be used to dilute samples, since poly(ethylene) does not show distinct absorption in the FIR region.

2.3.3   Features of Diffuse Reflection Spectra
When the diffuse reflection measurement contains a higher reflection component, the observed spectrum will have more distortion due to overlap of the reflection component.   Thus, removal of the reflection component from the diffuse reflection measurement will improve the quality of the diffuse reflection spectrum.

In order to achieve this[23]:
(i)   The sample must be ground as fine as possible.
Consider a sample composed of large particles, first.   In this case each particle of the sample

is likely to have well-developed crystalline faces and the specular reflection component becomes relatively significant.   In addition, the larger size particles absorb the radiation almost completely, leaving only a fraction of the transmitted radiation in the diffusely reflected radiation. On the other hand, when the powder size is small, for example as small as or smaller than the wavelength of the IR radiation, the contribution due to specular reflection at particle surfaces becomes small because the packing of the particles becomes more efficient, and for the IR radiation the system is quasi-continuous.

Thus, repeated refractions become more dominant and absorption component becomes more important.   However, the specular reflection at the first surface of the sample will not be removed, leaving the distortion in the diffuse reflection spectrum.

(ii) The sample must be diluted in KCl, KBr or other non-absorbing fine powder.

The density of the sample particles becomes small when the sample is diluted with an optically inert medium.   Thus, the specular reflection component from the first layer of the sample will be largely suppressed.   In addition, the specular reflection by the sample particle at the air-gap interface will be largely replaced by the specular reflection at the boundary between sample particles and diluent.   In this case the reflectance at the boundary between sample and diluent becomes almost 0, since the refractive index of KCl or KBr is closer to the average value of the sample refractive index than air.

(iii) The optical design must be arranged so that the specular reflection component is prevented from reaching the detector.

There are two types of optical arrangements in commercially available diffuse reflection accessories.   In one of the optical designs, the collimating mirror for the incident radiation, sample cup, and mirror collecting the reflected radiation are aligned along a line.   In this case, there is an inherent possibility of collecting specular reflection from the sample and sample cup. Another arrangement has the collecting mirror off the line, reducing the possibility of collecting the reflected radiation from the cup.   The user must be careful to adjust the orientation of the mirrors even for the off-line design in order to avoid collecting the specular reflection component from the sample cup.

Based on the recommendations by many researchers,[23] the rules of thumb are:

(*i*) The Kubelka-Munk corrected diffuse reflection spectrum (*vide infra*) of most organic samples diluted in finely grounded KCl or KBr below *ca.* 5wt% will be comparable with the transmission spectrum (in absorbance unit) taken with the KBr pellet method.   However, the particle sizes of the sample and KCl or KBr particles will influence the quality of the spectra.   It is reported that the particle size should be smaller than a few micrometers.

(*ii*) In the case of inorganic compounds, which show quite strong absorption bands like the $1,100 \text{ cm}^{-1}$ band of quartz and the $1,050 \text{ cm}^{-1}$ band of the $SO_4^=$ ion, the sample must be grounded very fine, for example less than a few tenth of micrometers in diameter and be diluted to *ca.* 0.5%. The reader will find examples of the influence of particle size and dilution in Part II of this volume.

Let us discuss the feature of diffuse reflection spectra.   The IR radiation whose wavelength corresponds to a weak absorption band can travel long distances by repeated refractions.   Thus, the diffuse reflection measurement makes the effective sample thickness larger for weak absorptions.   On the other hand, IR radiation corresponding to positions of strong peak will be totally absorbed by the long effective pathlength, while the IR radiation after passing through the very shallow surface of the sample with a limited number of refractions will survive and be detected.   In other words, the diffuse reflection scheme makes the effective sample thickness small for the strong bands.   So it is clear that the peak intensity of a diffuse reflection spectrum is not linear with sample concentration.   In order to correlate the diffuse reflection spectrum to

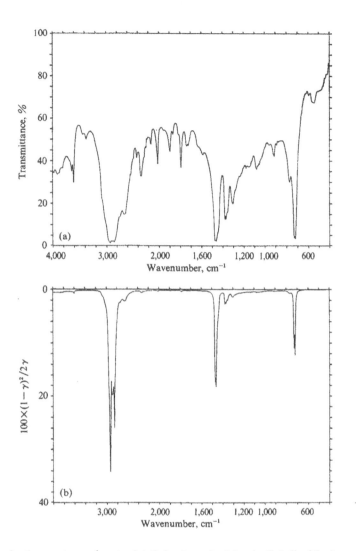

Fig. 2.17   Diffuse reflection spectrum of neat poly(ethylene) powder (a) and a Kubelka-Munk corrected spectrum (b).

the concentration, the so-called Kubelka-Munk calculation[24] will be needed, particularly if strong absorptions are to be used as the analytical bands.   Kubelka-Munk unit $f(\tilde{v})$ is given by

$$f(\tilde{v}) = 2.303a(\tilde{v})/s(\tilde{v}) = (1 - R(\tilde{v}))^2/2R(\tilde{v}) \qquad (2.20)$$

where $a(\tilde{v})$, $s(\tilde{v})$, and $R(\tilde{v})$ are absorbance, scattering coefficient, and diffuse reflection at frequency $\tilde{v}$, respectively.   Eq. 2.20 indicates that the Kubelka-Munk spectrum is comparable to an absorption spectrum in absorbance units. $R(\tilde{v})$ must be carefully measured using a diluted sample as mentioned in the experimental procedure Section 2.3.2, because the Kubelka-Munk treatment cannot remove the distortion due to the reflection component.   Fig. 2.17 shows a pair of diffuse reflection and Kubelka-Munk spectra of neat poly(ethylene) powder.   As expected, the Kubelka-Munk spectrum, Fig. 2.17(b), agrees well with the absorption spectrum of the same polyethylene except the CH stretching vibration at 2,900 – 2,800 cm^{-1}.   The discrepancy in this region is due to the specular reflection component, which is not accounted for by the Kubelka-Munk treatment.   On the other hand, diffuse reflection spectrum, Fig. 2.17(a), shows many peaks which are quite weak in the absorption spectrum; this is consistent with the previous statement that diffuse reflection measurements apparently enhance the intensities of weak absorptions.

2.4   Photo-Acoustic Detector

2.4.1   Principle

When a substance absorbs IR radiation, transition from its vibrational ground state to an excited state occurs.   The vibrational excited state with higher energy relaxes into the ground state immediately by converting this excess energy to *thermal energy*.[25]   Thus, due to absorption of IR radiation and subsequent thermal relaxation, the substance is heated.   The thermal wave thus generated diffuses through the substance and reaches the surface of the substance.

The heat which reaches the surface will further transfer to the surrounding gas, expanding the gas.   The amount of generated heat and subsequent expansion of the gas depend upon the amount of the absorbed radiation.   As a result, the gas pressure changes with frequency F, when the sample which is placed in the sealed vessel filled with inert gas is irradiated with IR radiation whose intensity is modulated with frequency F.   A microphone inside this sealed vessel detects this pressure change.[26]   In other words, the device described here can be used as a detector to monitor the absorption of the light.   In Fig. 2.18, the absorption of monochromatic radiation and the diffusion of the thermal wave is illustrated together with the design of a PA-detector. This effect is not limited to IR light but also occurs with ultraviolet and visible radiation and the spectrometry using this phenomenon is called *photo-acoustic spectrometry* (PAS).   Remember that

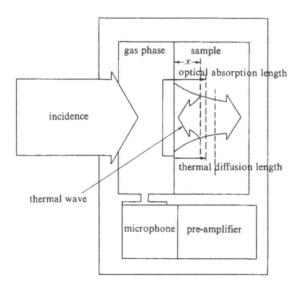

Fig. 2.18   Conceptual diagram showing absorption of IR radiation followed by a transfer of the heat wave and generation and detection of the sound.

the intensity of the IR radiation at the FT-IR sample compartment is modulated as shown in Fig. 1.5, the modulation frequency of the radiation with frequency $\tilde{v}$ being given as $f = F\tilde{v}$ (Hz) (Eq. 1.1) as shown in Chapter 1.   The output signal of the microphone is therefore similar to the interferogram of the incident IR radiation *times* the sample absorption, because the heat wave generated in the inert gas is modulated with the same frequency.   The Fourier transform of the signal will give us the single beam spectrum of the FT-IR system overlapped with the extra signal generated by sample absorption.

Since the mechanism of sound generation is dependent on both the light absorption and heat diffusion processes as described above, the PA spectrum has the nature of an absorption spectrum modified by the thermal diffusion characteristics of the sample.   It is, therefore, worthwhile to see what happens to the IR radiation and the thermal wave in the sample.   Let us define the absorption coefficient of the sample as $\beta(\tilde{v})$ (in cm^{-1} units).   Absorption of the radiation follows the Lambert-Beer law,

$$I = I_0 \exp(-\beta(\tilde{v})x) \tag{2.19}$$

where x is the distance from the surface of the sample (see Fig. 2.18).   Therefore, the intensity of the incident radiation becomes $1/e$ times of the initial intensity at distance $1/\beta(\tilde{v})$, *because the sample absorbs the incident radiation*.   Thus, *the optical absorption length*, $r_0 = 1/\beta(\tilde{v})$, is a good indicator of the distance that incident IR radiation can travel through the sample.

The conduction of the thermal wave is expressed by the classical diffusion equation[27] as

$$(\delta u/\delta t) = \kappa(\delta^2 u/\delta x^2) \qquad (\kappa > 0, \quad t < 0, \quad 0 < x < \infty) \tag{2.20}$$

for the one-dimensional case with a thermal diffusivity κ (cm^2/sec), where u, x, and t are temperature, distance, and time.   In order to estimate how far the thermal wave can travel through the sample, let us consider a bar of infinite length.   When the temperature at one end of the bar changes sinusoidally due to heat generation, *i.e.*, $u(0, t) = u_0 e^{i\omega t}$, the temperature of the bar at the position x at time t is given by the following[28]:

$$u(x, t) = u_0 \exp\{-(\omega/2\kappa)^{1/2}x\}\sin\{\omega t - x(\omega/2\kappa)^{1/2}\}* \tag{2.21}$$

The exponential term of Eq. 2.21 describes the decay of the thermal wave and $r_t = 1/((\omega/2\kappa)^{1/2})$ is the distance at which the thermal wave decays to $1/e$, the analogy of deriving r_0 for the decay of the incident IR radiation mentioned above.   Thus,

$$r_t = 1/((\omega/2\kappa)^{1/2}) = (2\kappa/\omega)^{1/2} = (2k/\omega\rho c)^{1/2} \tag{2.22}$$

gives us the distance into the substance which contributes to the PA-FTIR spectrum.   This is shown in Fig. 2.18 as vertical lines, and is called the *thermal diffusion length*.   Here terms K, ρ, ω, and c are thermal conductivity (cal cm^{-1} s^{-1} °C^{-1}), density (g cm^{-3}) modulation frequency of incident radiation as given by Eq. 1.1, and specific heat (cal g^{-1} deg^{-1}), respectively.

Since the mechanism of the signal generation is given, let us describe the fundamental nature of the PA-FTIR signal.   To a good approximation, the PA-FTIR signal intensity, I, is given under certain circumstances by

$$I \propto I_0 \beta(\tilde{v})/\tilde{\omega} \tag{2.23}$$

where I_0, $\beta(\tilde{v})$, and $\tilde{\omega}$ are intensity of the incident radiation, absorption coefficient of the sample,

---

* This is the same equation of a famous problem for undergraduate students of physics to estimate the temperature of the ground at position x at time t when the surface temperature changes periodically in a day or a year.   Those who wish to understand the theory of PA spectroscopy should study the diffusion equation.

and the angular frequency of the modulation to light, respectively.   The interpretation of this equation indicates that the PA signal I is close to 0 where there is no absorption, *i.e.*, the reverse of what is observed in the usual absorption measurements.   Therefore, it is clear that the PA-FTIR signal gives us an emissivity type spectrum in *absorbance units*, because I is proportional to $\beta(\tilde{v})$.   It is also clear that the single beam PA-FTIR signal contains the FT-IR system spectrum characteristics, since I is proportional to I_0.   The term $\tilde{\omega}$ which contributes to I in Eq. 2.22 is a product of the mirror speed V and IR frequency $\tilde{v}$ and is described as $\tilde{\omega} = 4\pi V\tilde{v}$ or $2\pi F\tilde{v}$ (F = OPD velocity).   Since the PA-signal is proportional to $1/\tilde{\omega}$, the PA-FTIR signal becomes weak when the mirror speed is fast and *vice versa*.   It is obvious from Eq. 2.23 that the low frequency signal is larger than the high frequency signal, because $\tilde{\omega}$ is small at low frequency or long wavelength, similar to the case of ATR spectra.*

It was also found that saturation of the PA-FTIR signal may occur with some highly absorbing samples such as carbon black, so that none of the spectral features of the sample is observed. Such a sample can be used as a background material for PA-FTIR spectroscopy, ratioing out the spectral feature of the FT-IR system from the single-beam PA-FTIR signal.   Thus, the PAS spectrum divided by the background spectrum for something like carbon black has the dimension of *absorbance but not %T*, because the *PA signal I* is proportional to $\beta(\tilde{v})$, as shown in Table 2.2.

TABLE 2.2   Classification of PA-signal Intensities

$r_0 > d$	(a)	$r_t > r_0 > d$	$I \propto I_0 \cdot \beta \cdot d \cdot \omega^{-1}$
	(b)	$r_t \sim r_0 > d$	$I \propto I_0 \cdot \beta \cdot d \cdot \omega^{-1}$
	(c)	$r_t < d;\quad r_t < r_0$	$I \propto I_0 \cdot \beta \cdot r_t \cdot \omega^{-3/2}$
$r_0 < d$	(d)	$r_t > r_0;\quad r_t > d$	$I \propto I_0 \cdot \omega^{-1}$
	(e)	$r_t < d;\quad r_t > r_0$	$I \propto I_0 \cdot \omega^{-1}$
	(f)	$r_t < d;\quad r_t < r_0$	$I \propto I_0 \cdot \beta \cdot r_t \cdot \omega^{-3/2}$

2.4.2   Experimental Procedure

It is recommended that the *slow* OPD velocity for background an sample observations be used because this yields the highest response.   Carbon black is usually used as a reference material to obtain the background spectrum.   PA-FTIR spectra of the sample are obtained by dividing the single beam spectrum of the sample in the sample cup by the background spectrum.   As explained above by Eq. 2.22, fast OPD velocities should be selected if the shallow surface of the sample is to be measured and slow OPD velocities for deep penetration into the sample.

PAS has quite high sensitivity for gas phase samples.   This implies that water vapor and CO_2 show strong PA signals.   Therefore, it is essential to purge the PA cell by dry gas in order to reduce the interference by these gases.   In addition, it is known that the intensity of the PA signal changes as the gas in the sample chamber changes.   By experience, it is recommended that helium gas be used to increase the PA signal.

The PA signal intensity depends on the pressure of the gas in the chamber.   Therefore, it is wise to wait a few minutes after purging and closing the cell before obtaining any spectrum to allow the PAS detector to achieve thermal equilibrium.

Sometimes a single beam spectrum is utilized to obtain spectra on small quantities of sample such a single fiber, or a paint chip.   In this case, besides the PA signal from the sample, the bronze sample cup creates only a very weak broad PA-signal, which can be subtracted with a PA-signal of the empty sample cup.   With such observation, each peak intensity includes both an

---

* This holds for continuous-scan type interferometers.   The step-scan type interferometer provides the opportunity to make thermal diffusion length independent of the wavelength of radiation.

absorption coefficient and intensity of incident radiation as clearly shown in Eq. 2.23. The reader will find such an example in Part II of this volume.

2.4.3 Features of the PA-FTIR Method

Rosencwaig and Gersho[29] explained the general features of the PA-FTIR signal more quantitatively, handling the thermal diffusion equation more precisely. Since the thermal diffusion equation for the PA-FTIR experiments is too complicated to provide a general solution, they have shown the solutions for six well-defined cases depending on the relative significance of the absorption coefficient $\beta(\tilde{v})$, thermal diffusion length r_o, and sample thickness d. When $r_o > d$, the sample is said to be "optically transparent," and is "optically opaque" when $r_o < d$. When $r_t > d$ the sample is "thermally thin" and "thermally thick" when $r_t < d$. Table 2.2 summarizes the six cases.

In cases (a) and (b), the entire thickness of the sample contributes to the absorption of radiation ($r_o > d$) and the entire thickness contributes to the PA-signal because $r_t > d$. Although the entire thickness contributes to light absorption in the case of (c), the sample to depth, r_t, contributes to the PA-signal because $r_t < d$.

Samples belonging to cases (d) and (e) do not show any spectral features and the sample may serve as the compound for generating the background spectrum. Since those samples have extraordinary strong absorption, absorbing 100% of IR radiation completely, the PS-signal shows only the optical characteristics of the FTIR system, and the PA-signal is considered to be saturated. In the case of (f), the sample again has strong absorption. However, if the thermal diffusion length r_t is shorter than the optical absorption length r_o, the sample will show spectral features. Therefore, even if the PA-signal does not show the sample absorption with a slow mirror speed, there is a chance to observe spectral features with higher mirror speeds.

Although optical and electrical processes such as the absorption of radiation and its electrical detection are instantaneous, the diffusion of the thermal wave is a slow process compared with the speed of light, and there is delay in the detection of the thermal wave. This causes a phase delay in the interferogram. McClelland[30] gave PA signal magnitude, $q_l(\lambda)$, and phase, $\psi_l(\lambda)$, for the sample whose thickness d is thicker than both r_t and r_o as follows:

$$q_l(\lambda) = \kappa(1 - R)(I_0(\lambda) \cdot \beta(\lambda)/\alpha s)(2/((\beta(\lambda)/\alpha s + 1)^2 + 1))^{1/2} \tag{2.24}$$

$$\psi_l(\lambda) = \tan^{-1}(1 + 2\alpha s/\beta(\lambda))) \tag{2.25}$$

where R is reflection of the radiation by the sample. Combining those two quantities into Eq. 2.26, McClelland showed that the linearity of the PA-FTIR spectra is extended significantly.

$$q_{ll}(\lambda) = q_l(\lambda)((\cot^2(\psi_l(\lambda) - 450) + 1)/2)^{1/2} \tag{2.26}$$

Those who are interested in obtaining PA-FTIR spectra with a linearity correction should consult with the FT-IR supplier, because it requires an algorithm corresponding to Eq. 2.26.

Reviewing the above-mentioned features of PA-FTIR spectra, the reader will find that cases (c) and (f) are the most interesting. By inputting thermal diffusivity α or thermal conductivity K, density ρ, and specific heat C as well as the OPD velocity and the frequency of interest in Eq. 2.22, one can set the observation depth artificially. For example, the thermal diffusivity α of a common polymer such as polyethylene is on the order of 1.3×10^{-3} cm^2 s^{-1}. Therefore, when the OPD velocity of 0.05 – 3 cm s^{-1} is employed, the thermal diffusion length for radiation at 1,500 cm^{-1} ranges from 24 to 3 micrometers. Compared to the ATR depth profiling where penetration depths range from 0.3 – 1.3 micrometers at the same frequency of the radiation using different IREs and incident angles, one can see that the PA-FTIR method is observing depths one order of magnitude larger than the ATR-FTIR method.

Provided with a series of PA-FTIR spectra taken with different r_t values, determination of the chemical composition as a function of the penetration depth can be done. A theoretical and experimental study[31] performed using the UV/VIS PAS technique showed that the *inverse Laplace transform* of such a series of spectra determined the chemical composition; the same *inverse Laplace transform* technique was discussed for ATR depth profiling in a preceding chapter.

The most prominent advantage claimed for PAS is that the PA spectra are independent of sample morphology.[32] For example, a sample of polystyrene in powder or pellet form, or having a rough surface, exhibits the same spectrum, although the signal intensity may change with morphology. This feature is compared with the fact that the diffuse reflection spectrum of a polystyrene pellet is unusable due to the dominating specular reflection component of the spectra.

Since the specular reflection component from the surface of a polystyrene pellet does not produce a thermal wave in the case of PAS, it is understandable that the PA-FTIR spectrum is not affected by the surface condition. However, since the reflection changes the absorption as given by Eq. 2.24, the reflection will cause a secondary effect to the PA-FTIR spectrum, that of reducing the overall signal intensity. In addition, it is not completely correct to assume that the PA-FTIR spectrum is not affected by sample morphology at all. In fact Burggraf and Leyden[31] have formulated the PA signal of a highly scattering sample. They have shown that PAS can be a reliable quantitative analysis technique, provided that the effects of acoustic background, light scattering and saturation effect are adequately accounted for.

Although the spectral subtraction method is quite useful for most applications using IR accessories, the PA signal is easily saturated and difference spectroscopy using PA-FTIR spectra is sometimes difficult. For example, an attempt to remove the water vapor spectrum from a PA-FTIR spectrum will be frustrated in most cases.

Teramae[34] showed that some film samples should be stacked firm to the bottom of the sample cup using two-sided adhesive tape. He showed experimentally and theoretically that the sound coming from the backside of the film will interfere with the sound coming from the front surface of the sample when there is a gap between the film and the bottom of the sample cup. These sounds will combine destructively when the backside thermal wave has the opposite phase to that of the front surface. Thus, some of the intense peaks lose intensity and sometimes a "ringing" artifact overlaps intense peaks of the PA-FTIR spectra, as shown by Fig. 2.19.

2.5 Fixed Angle and Variable Angle Specular Reflection Accessories

2.5.1 Principle

When the sample does not transmit IR radiation because the sample is too thick or the sample is coated on metal plate, the reflected radiation from the sample is the only means to obtain sample information. As shown in Fig. 2.20, radiation incident to the specularly reflecting sample with angle θ is reflected with the same angle and this reflected radiation carries the sample information. Accessories with a fixed incident angle and a variable incident angle are commercially available. For these accessories, a gold or aluminum mirror is used as the reference material because the reflectance of those mirrors is high ($>91\%$) and almost constant over the entire IR frequency range. The observed reflectance is not an absolute reflectance but a relative reflectance, the value being relative to the reference mirror.* The spectrum of a sample

---

* Those who wish to obtain the absolute reflectance must use an absolute reflectance accessory. Harrick Scientific supplies this accessory.

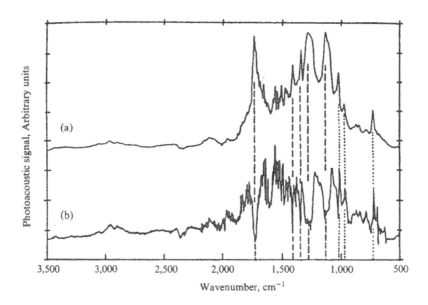

Fig. 2.19   PA-FTIR spectrum of polycarbonate film placed loosely in the sample cup, leaving an air gap between the sample and the sample cup (b) compared with a normal PA-FTIR spectrum (a).

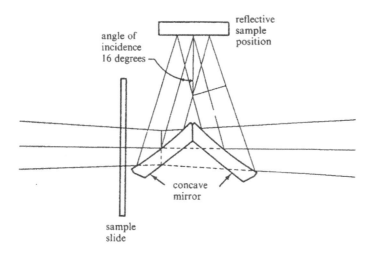

Fig. 2.20   Optical diagram of a fixed angle specular reflection accessory.

may be measured when the mirror is placed behind the sample so that the combination of the sample and the mirror is optically equivalent to the metal plate coated with a thin film.

2.5.2   Specular Reflection Accessory

A variety of "fixed angle specular reflection" accessories are commercially available.   These differ in the incident angle, varying from 6.5° to 80°.   The accessories with high incident angle, for example, the FT-80 from Spectra-Tech with an angle of incidence of 80° and GRA series from Harrick Scientific, are designed for the observation of thin layers on metal surfaces such as the lubricant on magnetic hard disk media, monomolecular films on metals, and barrier type oxide layers on metals.   An incident angle smaller than 20° is best suited to convert the specular reflection spectra into absorption spectra with the aid of the Kramers-Kronig integration (*vide infra*).   An optical diagram of one of the fixed angle specular reflection accessories is shown in Fig. 2.20.   The inident angle of this accesory is 16°.

A "variable angle specular reflection" accessory has a wide range of incident angles, e.g., 15° – 85°.   For example, Fig. 2.21 describes the optical design of the Harrick Scientific Model VRA-RMA.   The incident angle is selected by rotation of the plate holding A and A'.   The variable angle reflection accessory provides for versatile applications in IR spectroscopy.   It is possible to determine both the refractive index and thickness of the sample by measuring the locations of the fringe patterns at two incident angles.   The high incident angle is suited for measurements of thin layers on the metal substrates as mentioned above.   The medium angle of incidence such as 35° – 65° is an important range of incident angle for the measurement of thin layers on the non-metallic substrate such as glass, metal oxides, and polymers.

In addition, the variable angle specular reflection accessory enables one to obtain the transmission spectrum of polymer film without interferance fringe pattern.   The "monolayer" accessory from SPECAC further enables work on monolayers on liquid surfaces such as protein monolayer on the surface of water, in addition to the above-mentioned applications.   Many of these applications require the use of an IR polarizer and the reader will find them discussed in Part II.

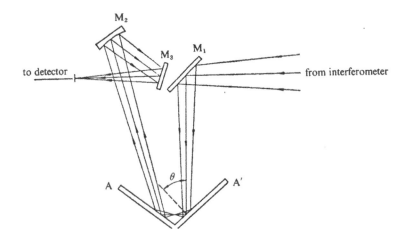

Fig. 2.21   Optical diagram of a variable angle reflection accessory.

2.5.3   Experimental Procedure
A.   Fixed angle specular reflection accessory

As shown in Fig. 2.20, a sample is placed at the focal position of the incident radiation, where a series of apertures with different diameters can be placed to match the size of the sample.   If there is no need to select a small area of the sample, the largest aperture is recommended for use because of the highest optical throughput.   The single beam spectrum with the reference mirror on the sample position is measured as the background spectrum.   The absorption spectra of film samples can be measured with this accessory.   In this case, place the film at the sample position and cover the film with the reference mirror.   This will give the absorption spectrum of the film with effective thickness about two times bigger than the actual thickness.

B.   Variable angle specular reflection accessory

The Harrick variable angle specular reflectance accessory VRA-RMA shown in Fig. 2.21 will be used to explain the procedure for a grazing angle reflection measurement.   This accessory specifies the highest incident angle as 75°.   Place an aluminum mirror at the A position; the A' position is for the reference mirror or sample.

When a incident angle higher than 75° is necessary, another accessory such as a "monolayer" accessory should be used.   However, in order to achieve such high incident angle, the reader must be aware that the incident radiation with a small beam diameter and a small divergence angle should be used.   Otherwise the incident radiation is directly collimated to the detector without any interaction with the sample.   The procedure to obtain a reflection spectrum is essentially the same as described earlier.

When a polarizer is used, set the polarization angle to the required experimental condition, referring to the angle marked on the indicator of the polarizer.   While only p-polarized light gives rise to spectral features from thin film on metal plate, both p- and s-polarized light will contribute to the spectral features of thin film on a nonmetal substrate, as will be discussed in Section 2.5.4.   In the case of thin film on metal plate, the polarizer must be set to 90° for the Harrick accessory, since the plane defined by the incident light and the reflected light lies horizontal.   However, if the reader cannot figure out which number on the marker should be used, 0 or 90, to provide p-polarization, follow the procedure described below.   Place the microscope cover glass at the A' position and set the incident angle to *ca.* 40° – 50° (close to the Brewster angle of the glass).   Rotate the polarizer element and check the energy level.   The setting to give the minimum energy level implies the polarizer is adjusted to give the parallel (p-)polarization.   The setting to give the maximum energy is for perpendicular (s-)polarization.

An incident angle of 80° – 85° or higher is recommended for grazing angle reflection measurement of thin layers (*vide infra*).   However, it will be far easier to start with a 75° or lower angle of incidence, because the length of the image at the sample position is $1/\cos(\theta)$ times the image in the sample compartment and a small misadjustment can easily ruin the experiment. Indeed, an angle of incidence of 70° to 75° will be adequate to observe thin organic material as thin as 2.5 nm.   Be careful that the mirror and the sample are free from any contamination, because the purpose of this experiment is to measure the IR spectrum of thin layer.   The reference mirror and sample must not be touched by bare fingers.   Wear rubber gloves or finger cots to avoid contamination from handling.

2.5.4   Features of Specular Reflection Spectroscopy
It should be remembered that the theoretical treatment of spectra on fixed angle and variable angle reflection accessories depends on the sample thickness and the presence or non-presence of backing material of the sample.   Four cases will be discussed.   The first three cases are samples

coated on a smooth metal substrate or backed by a mirror.   The fourth case is the measurement of non-metallic substrates coated with thin layer such as surface coated glass or polymers.

(i)   When the sample is a thin film of thickness from a few to a few tens of micrometers, the reflection spectrum becomes an absorption spectrum of the thin film with effective thickness twice that of the actual thickness when the incident angle is near normal.   This is because the radiation incident to the sample passes through the film and is reflected by a mirror or metal surface and again passes through the sample.   Since a few percent of the radiation is reflected by the surface of the film, the reflected radiation and the radiation through the sample can interact to cause, for example, an interference fringe pattern in the absorption spectrum.   In fact the reflected rays from the surface cause a zero offset and sometimes a distortion in the peak shapes of strong bands near the peak just as would be expected if a true specular reflection measurement were made for a dielectric film.   One can calculate the film thickness using the spacing of the fringe pattern, using the following equation:

$$d \, (\text{cm}) = \Delta m / \{ 2(\tilde{v}_1 - \tilde{v}_2)(n^2 - \sin^2 \theta)^{1/2} \} \tag{2.27}$$

where $\tilde{v}_1$ and $\tilde{v}_2$ are frequencies in cm^{-1} units of the first and the $\Delta m + 1$ peak (or valley) of the fringe pattern.   Terms n, Δm, and θ are the refractive index of the film, the number of valleys (or peaks) between $\tilde{v}_1$ and $\tilde{v}_2$ and the incident angle, respectively.   Eq. 2.27 indicates that, by using two incident angles with the variable angle reflection accessory, it is possible to determine thickness and refractive index of the sample as mentioned above.

(ii)   When the thickness of the film on the metal plate becomes very thin compared with the wavelength of the radiation ($d \ll \lambda$), the observation of the spectrum will be frustrated with normal incidence. An observation with parallel polarization at a grazing angle incidence such as 75° –85° will increase the intensity of the absorption type signal substantially.   Note that this enhancement occurs only when the thin film (such as a monomolecular layer) is coated on a metal plate or certain semiconducting plates.   Fig. 2.22 shows the electric fields of incident and reflected radiations at the surface of the metal.

In the case of perpendicular polarizations (a) and (b), the electric fields of the incident and reflected radiation cancel each other regardless of the incident angle, because the phase of the radiation changes 90° due to reflection by the metal.   Thus, there is no interaction between substance and radiation, giving no change in signal.   In the case of parallel polarization with normal incidence, (c) is the same as (b).   Only parallel polarization with a grazing angle of incidence, (d), will provide a strong stationary electric field at the metal-film interface that can cause a strong enough interaction between the substance and radiation to give a change in signal. Furthermore, it has been demonstrated[35] that the signal intensity becomes maximum at 80° – 85°.

Theoretical and experimental development of this technique[35] is attributed to Greenler, McIntyre and Suetaka.   Suetaka[36] showed that the relative reflectance change of parallel polarized light (//) due to the presence of a thin film on the metal is expressed as

$$(\Delta R / R_0)_{/\!/} = -16\pi n_1^3 n_2 k_2 \sin^2 \theta d / \{ (n_2^2 + k_2^2)^2 \lambda \cos \theta \} \tag{2.28}$$

where $\Delta R = R_d - R_0$ is the difference in reflectance with (R_d) and without (R_0) the thin film. n_1 is the refractive index of the medium from which the incident radiation comes to the film (if it is air, $n_1 = 1.0$). n_2 and k_2 are the real and imaginary parts of the complex refractive index of the film material, respectively.   Terms, λ, θ, and d are the wavelength of the radiation, the incident angle, and the thickness of the film, respectively.   As it is clearly expressed by Eq. (2.28), the grazing angle reflection spectrum of a thin film on a metal plate is *not* an absorption spectrum. However, when $k_2^2 \ll n_2^2$ (this is a case when the film is composed of an organic compound),

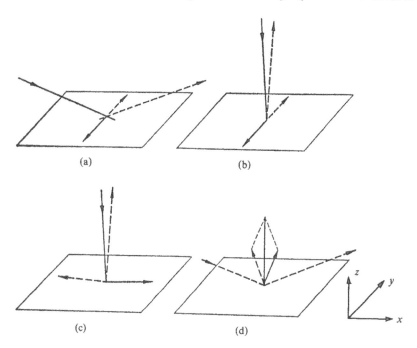

Fig. 2.22   Electric fields at a metal surface.   (a) grazing angle incidence of s-polarized light.   (b) normal incidence of s-polarized light.   (c) normal incidence of p-polarized light.   (d) grazing angle incidence of p-polarization generating a strong electric field normal to the metal surface.

Eq. (2.28) reduces to a simpler form

$$(\Delta R/R_0)_{//} = -(4n_1^3 \sin^2 \theta / n_2^3 \cos \theta)\alpha d \qquad (2.29)$$

where $\alpha (= 4\pi k_2/\lambda)$ is the absorption coefficient of the film material.   Thus the grazing angle reflection spectrum of an organic compound on a metal plate becomes an absorption-like spectrum.   The term in parentheses in Eq. 2.29 indicates the sensitivity gain by this method.   As explained in Section 2.2 for the ATR technique, an anomalous dispersion of n_2 will cause a shift in peak position.   Eq. 2.29 suggests that a shift in band position towards higher frequencies with occur because n_2^3 exists in the denominator of Eq. 2.29.   It is well known that inorganic species on metal show absorption-like spectra with a sizable high frequency shift (blue shift) compared with absorption spectra.[37]   The shift described in Eq. 2.28 was recently explained quantitatively by a computer simulation method.[38]   The reader will see an example of the high frequency shift even in the case of an organic compound in Part II (6.3D).   This high frequency shift caused by the nature of the grazing angle reflection technique is opposite to the low frequency shift observed in the ATR technique.

   It must be understood that the last two equations hold only when the thin film is backed by a *metal*.

(iii)   When the film is thick, the radiation which is refracted into the sample is absorbed completely and does not come out.   However, a few percent of the radiation is reflected at the surface and the radiation reflected at the surface shows spectral features.   The spectrum thus observed is called *"specular reflection spectrum."*   In this case the presencce of the metal plate makes no difference from a thick sample without a metal plate.

Specular reflection spectra look different from absorption spectra.   Fig. 2.23(a) is a specular reflection spectrum of a 4 mm thick smooth plate of poly(methylmethacrylate), PMMA.   The spectrum looks like a derivative type making qualitative and quantitative analyes very difficult. Fortunately, the specular reflection spectrum is convertible to the absorption spectrum by computer calculation, as shown in Fig. 2.23(b).   The process is known as Kramers-Kronig analysis,[39] which determines the real and the imaginary parts of the refractive index, n and k, from either a reflection or absorption spectrum (Fig. 2.23(c) and (d)).   Since both reflection and absorption spectra are expressed in terms of n and k, those spectra are interconvertible through Kramers-Kronig analysis.   Moreover, it is possible to calculate the grazing angle reflection spectrum of a thin film on a metal plate from either a reflection or absorption spectrum measured as explained in a previous Section, 2.5.4. (*ii*)   As the theory of Kramers-Kronig analysis reveals when calculating an absorption spectrum from a specular reflection spectrum, it is much simple to use the specular reflection spectrum measured with normal incidence.   Therefore, the reflection accessory with a smaller incident angle is more suitable for Kramers-Kronig analysis. It is reported that the incident angle should be smaller than 20°.[40]

(iv)   When a non-metallic substrate such as a polymer, metal oxide, or glass is coated with thin layer, both the fixed angle and variable angle specular reflection accessories are used to obtain information of the surface.   If we classify this kind of material as one of the examples of case (*iii*), we would expect the treatment of the specular reflection spectra based on Kramers-Kronig analysis to yield the absorption spectrum of the layered substance.   Indeed, as will be shown in Part II (Section 11-4B), Kramers-Kronig analysis of the surface-coated material and uncoated reference material will provide the opportunity to obtain the difference spectrum showing the surface-coating material.   On the other hand, if the surface layer is very thin, the variable angle specular reflection accessory with polarizer may be used.   Theoretical treatments using the Fresnel reflection coefficients for the layer structure indicate the incident angle which should be used to optimize the signal from the surface material for p- and s-polarized observations.   Those who wish to work on surface-coated nonmetallic substrates are advised to read the article by McIntyre regarding stratified medium model.[41]   Readers will find examples in Part 2 (6-2D, E, F).

Whether the sample thickness is small or large as mentioned in (*i*) and (*iii*) is relative to the absorption coefficient.   For instance, a 1-mm-thick polymer film on a metal plate belongs to category (*iii*) for mid-IR observation, because the sample is too thick, *i.e.*, nearly totally absorbing in the mid-IR region.   However, the same sample has adequate thickness in the near-IR region for category (i), because the absorption in the near-IR region is about two orders of magnitude weaker than in the mid-IR region.   Fig. 2.24 is a specular reflection spectrum of a 1-mm-thick polycarbonate plate backed by an aluminum mirror.   In the near-IR (7,000 – 4,000) region, the spectrum shows a typical *absorption spectrum* in %T units.   However, in the mid-IR (2,000 – 450 cm^{-1}) region, a typical *specular reflection spectrum* is observed, because all of the refracted radiation is absorbed by the sample and only the reflected radiation is seen by the detector.   The inserted spectrum in Fig. 2.24 is an absorption spectrum obtained through Kramers-Kronig analysis of Fig. 2.24.   The region from 4,000 to 2,000 cm^{-1} is not easily utilized, because the radiation is totally absorbed in some regions and not in others.

Finally, it is worth mentioning the technical names of the specular reflection methods. Method (*i*) is often called "Infrared Reflection-Absorption Spectroscopy, or IR-RAS." Method (*ii*) is also called "Infrared Reflection-Absorption Spectroscopy, or IR-RAS" and is sometimes called "Grazing Angle Reflection Spectroscopy" or "High Sensitivity Reflection Spectroscopy."   Methods (*iii*) and (*iv*) are usually called "Specular Reflection spectroscopy," "External Reflection Spectroscopy," or "Reflection Spectroscopy."

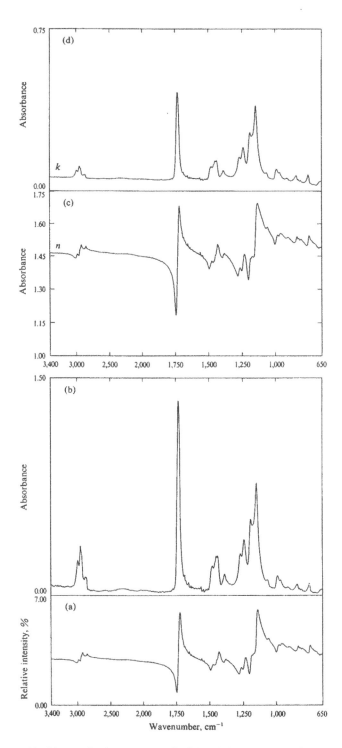

Fig. 2.23 (a) Near normal incidence reflection spectrum of poly(methylmethacrylate) plate.   (b) Absorption spectrum calculated through Kramers-Kronig integration.   (c) *n*-index and (d) *k*-index of poly(methylmethacrylate) calculated through Kramers-Kronig integration.   Spectrum (a) was used to calculate spectra (b)–(d).

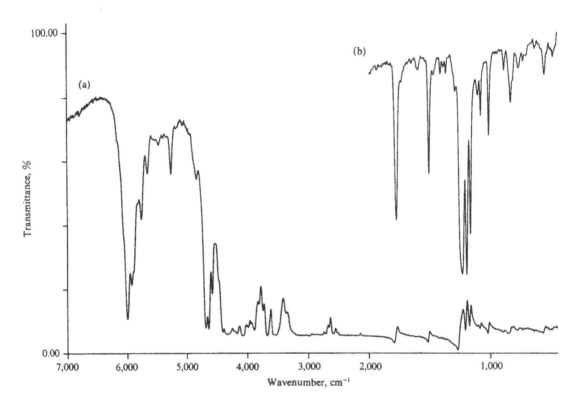

Fig. 2.24 Specular reflection spectrum of a 1-mm-thick polycarbonate plate backed by Al mirror (a) and a mid-IR absorption spectrum calculated from a part of (a) through Kramers-Kronig integration (b).

2.6 Emission Accessory

2.6.1 Principle

Although emission spectroscopy using an FT-IR spectrometer proved to be one of the advantages[42] of FT-IR over the traditional dispersive type spectrometer, emission measurements are not so common in infrared analysis. There are several reasons for this. First of all, the emission method requires a dedicated sample compartment, rendering the system expensive. Second, a good black-body source is required as a reference. Third, since the emission spectrum is taken based on the assumption that the temperatures of the sample and the detector are different, any optical component in the system such as the beam-splitter and mirrors can become a source of stray light, and because these components may also have different temperatures, the quantitative analysis will become difficult or tedious.*

However, if the sample cannot be placed in any of the transmission cells or IR accessories, the emission method may be the only available technique for IR analysis. In this section, a traditional method is described to observe emission spectra of heated materials using the FT-IR spectrometer with a dedicated emission compartment, together with a convenient method

---

* A cooled sample can produce an emission spectrum. In this volume , however, only heated samples are discussed.

without a dedicated emission compartment.

The heated sample replaces the IR source of the FT-IR system in emission measurements. Therefore, for emission measurements, the optical path is switched so that the heated material placed in the dedicated emission sample compartment has an optical position equivalent to the IR source.

An alternate optical arrangement using a standard sample compartment may also be employed. The reader may recall that half the radiation from the source to the interferometer is reflected back to the source by the interferometer. Taking advantage of this nature of the interferometer, Spragg and Ford[43] demonstrated that radiation from the heated sample which is introduced to the interferometer by placing a half-mirror in the sample compartment comes back to the sample compartment after being modulated in the interferometer. This arrangement provides emission spectra intense enough for qualitative analysis, although the efficiency of the system is at most 50% that of the dedicated sample compartment.

Regardless of whether a dedicated emission sample compartment or a sample space accessory is used, single beam spectra, $I_s(t)$, $I_s(0)$, $I_r(t)$, and $I_r(0)$ should be measured, where the suffix, r and s, indicate reference and sample, respectively and t and 0 in parentheses show the sample temperature and ambient temperature, respectively. Using these data, the emissivity of the sample is given by Eq. 2.30.

$$E(\tilde{v}) = (I_s(t) - I_s(0))/(I_r(t) - I_r(0)) \tag{2.30}$$

2.6.2 Emission Accessory

In order to perform the emission experiment, it is necessary to use a heated black body accessory to provide the reference emissivity. This accessory will also be used to heat the sample mounted at the sample position as the reference black body, so that the optical path for the reference and the sample are same. However, the sample space emission accessory is needed when the FT-IR system is not equipped with a dedicated sample compartment for the emission measurement. Harrick Scientifics supplies a microscopic emission accessory providing the capability to measure the emission spectrum of a sample as small as 0.5 mm in diameter. The Connecticut Instrument Company supplies the sample space emission accessory with the black body furnace for general observation as well as a special furnace with goniometer.

It was found[44] that a metal piece coated with thick black paint such as High Temperature Flat Black Paint (Zynolyte, Carson, California) works as a good reference material.

2.6.3. Experimental Procedure

A. Using a dedicated emission sample compartment

When the DTGS detector is used, the $I_s(0)$ and $I_r(0)$ are both nearly nil, because the temperature of the detector is the same as that of the reference or sample. Thus, emissivity is approximately given by

$$E(v) = I_s(t)/I_r(t) \tag{2.31}$$

Measure the single beam emission spectrum of the heated reference material as a background spectrum, followed by that of the heated sample. The ratio of those spectra gives the emission spectrum of the sample as shown in Eq. 2.31.

When an MCT detector is used, the room temperature spectrum of each reference and sample must be measured and Eq. 2.30 must be used instead of Eq. 2.31, because any optical component in an interferometer becomes an IR source.

B.   Using a sample space emission accessory

When using a sample space emission accessory, the IR source must be turned off and observation postponed until the source has cooled to room temperature. Although the experimental procedure using a DTGS detector is more or less the same as in the case of the dedicated sample compartment, the use of the MCT with a sample space emission accessory necessitates special precautions. Since the emission from the internal IR source at room temperature is detected by the MCT detector, it becomes the room temperature emission source. In addition, the interferogram of the internal IR source emission is 180° out of phase from that of the sample. Therefore, Fourier transform emission spectra calculated with the normal phase correction procedures may not cancel the room temperature emission component. As a result subtraction of the room temperature reference and sample spectra must be done with interferograms. Since the efficiency of the sample space emission accessory is not as good as the efficiency of the interferometer to the IR source, the amplitude of the interferogram due to the room temperature IR source can be larger than that from the sample. If this is the case, the energy value or amplitude of the centerburst will be minimum when the accessory is adjusted for the maximum throughput.

2.6.4   Features of Emission Spectroscopy

The broad fields of application cover coating technology and surface characterization of solid samples. For instance, emission spectra are frequently investigated in relation to the oxidation of metal surfaces.[45] Also, thermal curing and degradation of polymers can be studied by the emission method. Surface treated solids can be studied, especially when the sample is bulky and difficult to place in the usual absorption and reflection accessories. One unique application is in the study of catalysis using a powder catalyst[46] at elevated temperatures.

When the emission technique is employed to study coating on metals, the thickness of the coating may affect the quality of emission spectra. For instance, when the thickness of the coating material is very much smaller than the wavelength of the emission radiation, a large viewing angle is necessary to obtain a strong emission, and no spectral feature may be observed when the normal viewing angle is employed. Note the similarity to the grazing angle reflection-absorption spectroscopy measurements. Radiation from a dipole moment normal to the metal surface and radiation reflected by the metal will always combine constructively in the grazing direction as shown in Fig. 2.22(d),[47] since the optical path difference is far smaller than the wavelength. Therefore, the emission from the dipole results in strong spectral features. On the other hand, the dipole moment has no component in the direction normal to the surface, leaving no spectral features in the normal direction. When the vibrational dipole moment lies parallel to the metal surface, the electric field from the dipole and the field reflected at the metal surface have phase difference approaching 180°, and two rays combine destructively to leave zero spectral features, regardless of the viewing angle. In analogy to the grazing angle reflection-absorption spectroscopy, the emission spectrum of a thin coating of a metal plate observed with a large viewing angle will show a high frequency shift resulting from the contribution of a real part of the refractive index of the coating material. These features are the mirror image to the grazing angle reflection-absorption spectroscopy of the thin coating of the metal plate.

The readers will find in Section 6-3C that the emission spectrum of the thin oxide layer of the aluminum plate shows an emission peak at 960 cm^{-1} with a large viewing angle, which is the exact mirror image to the grazing angle reflection-absorption spectrum.

When the sample on the metal plate is thick, the viewing angle will not make much difference in the emission spectrum. Nor will the presence of a metal plate affect the spectrum. However, when the sample is heated in an imperfect furnace where the temperature of the entire sample is not the same, the sample thickness will cause significant effects on the emission spectrum.   Since the sample is often heated from the back side of the sample, the surface temperature will be lower than rest of the sample.   Thus, the emission radiation coming from the inside of the material may be absorbed by the lower temperature portion of the sample at the surface.   This phenomenon, a reabsorption of the emission spectrum, causes the inversion of the peak shape around the peak top.   When the sample thickness is larger, the effect becomes bigger and the reabsorption will eliminate spectral features.[48]

Carbon-filled samples also show reabsorption or inversion of the emission spectrum.   In this case, carbon black acts as an IR source and the blackbody-like emission from carbon particles is absorbed by the surrounding material, resulting in a weak absorption type spectrum.[49]

NOTES AND REFERENCES

1. (a) ATR Cassegrain for IR-PLAN by Spectra-Tech. Inc.;
    (b) Split Pea ATR microsampling accessory by Harrick Scientific.
2. Grazing Angle Cassegrain for IR-PLAN by Spectra-Tech. Inc.
3. C. W. Mason, *Handbook of Chemical Microscopy*, Vol. 1 (4th edition), Wiley, New York (1983).
4. R. D. Fraser, *J. Chem. Phys.*, **21**, 1511 (1963).
5. Courtesy of Spectra-Tech Inc.
6. Eugene Hecht, *Optics*, 2nd edition, Chapter 10, Addison-Wesley Pub., Massachusetts (1987).
7. (a) Chapter 2 of reference (6); H. Born and E. Wolf, *Principle of Optics*, 2nd ed., Pergamon Press, New York, (1964).
8. R. G. Messrschmidt, "Minimizing Optical Nonlinearities in Infrared Microspectroscopy", in *Infrared Micro-spectroscopy*, Practical Spectroscopy series, volume 6, page 1 (1988) Marcel Dekker, (New York and Basel).
9. A. Hanning, *Appl. Spectr.*, **42**, 90 (1988).
10. reference (9) also see B. Chase in reference (8) for other sources of corrections.
11. The present authors are deeply indebted to Dr. Reikichi Iwamoto of Osaka Institute, MITI, to let us summarize the chapter on the ATR technique from the book, reference (12b), written by one of the present author (K. N.) and Dr. R. Iwamoto.
12. (a) *Internal Reflection Spectroscopy* (3rd Printing), (ed. N. J. Harrick), Harrick Scientific Corp., Ossining, New York (1987);
    (b) K. Nishikida, R. Iwamoto, *Material Analysis by Infrared Spectroscopy* (in Japanese).   Kodansha, Tokyo (1988);
    (c) F. M. Mirabella, Jr (ed.), *Internal Reflection Spectroscopy*, Mercel Dekker, New York.
13. K. Ohta, R. Iwamoto, *Appl. Spectr.*, **39**, 418 (1985).
14. ref. 12(b) pp. 128 – 134.
15. ref. 12(b) pp. 122 – 128.
16. T. Hirshfeld, *Appl. Spectr.*, **31**, 289 (1977).
17. D. J. Carlsson, D. M. Wiles, *Macromolecules*, **4**, 174 (1971).
18. (a) L. J. Fina, G. Chen, *Vibrational Spectroscopy*, **1**, 353 (1991),
    (b) L. J. Fina, G. Chen, J. E. Valentini, *Ind. Eng. Chem. Res.*, **31**, 1659 (1992).
19. ref. 12(a) chapter 7.
20. J. E. Olsen, F. Shimura, *Appl. Phys. Lett.*, **53**, 1934 (1988).
21. (a) P. Kubelka and F. Munk, *Z. Tech. Phys.*, **12**, 593 (1931);
    (b) G. Kortum and H. Schlttler, *Z. Elektrochem.*, **57**, 353 (1953);
    (c) G. Kortum and D. Oelkrug, *Z. Naturforsch.*, **19A**, 28 (1964).
22. M. Fuller, *Dissertation*, Ohio University (1980).
23. (a) P. W. Yang, H. M. Mantsch, F. Baudais, *Appl. Spectr.*, **40**, 974 (1986);
    (b) P. J. Brimmer, P. R. Griffith, *Appl. Spectr.*, **41**, 791 (1987).
24. P. Kubelka, F. Munk, *Z. Tech. Phys.*, **47**, 64 (1931).
25. Lifetime of vibrational excited states is typically 1 – 10 pico-second.
26. A. Rosencwaig, *Opt. Comm.*, **1**, 305 (1973).
27. W. J. Moore, *Physical Chemistry*, 3rd edition, Chapter 9, Prentice-Hall, New York (1962).
28. E. W. Swokowski, *Calculus*, Prindle, Weber, & Schmidt, Boston (1976).
29. (a) A. Rosencwaig, A. Gersho, *J. Appl. Phys.*, **47**, 64 (1976).
30. J. F. McClelland, *Anal. Chem.*, **55**, 90A (1983).
31. L. W. Burggraf, D. E. Leyden, *Anal. Chem.*, **53**, 759 (1981).
32. A. Harata, T. Sawada, *J. Appl. Phys.*, **65**, 959 (1986).
33. D. W. Vidrine, Photoacoustic Fourier Transform Infrared Spectroscopy of Solids and Liquids, in *Fourier*

Transform Infrared Spectroscopy, Volume 3, (ed. by J. R. Ferraro and L. J. Basele), Academic Press, New York (1982).

34. N. Teramae, S. Tanaka, *Anal. Chem.,* **57,** 95 (1985).
35. (a) R. G. Greenler, *I. Phys. Chem.,* **44,** 310 (1966);
     (b) S. A. Francis, A. H. Ellison, *J. Opt. Soc. Amer.,* **49,** 131 (1959);
     (c) J. D. E. McIntyre, D. E. Aspnes, *Surf. Sci.,* **24,** 417 (1971);
     (d) W. Suetaka, *Bunkō Kenkyu,* **26,** 251 (1977).
36. ref. 35(d).
37. G. W. Poling, *J. Electrochem. Soc.,* **116,** 958 (1969).
38. K. Nishikida, R. W. Hannah, *Appl. Spectr.,* **46,** 999 (1992).
39. (a) D. M. Roessler, *Brit. J. Appl. Phys.,* **17,** 1313 (1966);
     (b) J. P. Haurane, P. Neelakantan, R. P. Young, R. J. Jones, *Spectrochim. Acta,* **32A,** 85 (1976).
40. H. Takahashi, J. Hiraishi, N. Ishii, *Bunkō Kenkyu,* **25,** 153 (1976).
41. (a) J. D. E. McIntyre, *Advances in Electrochemistry and Electrochemical Enginerring,* **9,** 61 (1973); also read;
     (b) D. L. Allara, R. G. Nuzzo, *Langmuir,* **1,** 45 (1985);
     (c) R. M. A. Azzam, N. M. Bashara, *Elipsometry and Polarized Light,* North-Holland Personal Library, Amsterdam (1977).
42. (a) J. M. Chalmers, M. W. Mackenzie, *Advances in Fourier Transform Infrared Spectroscopy,* (ed. M. W. Mackenzie), pp. 170 – 188, Wiley, New York (1988);
     (b) B. Chase, *Appl. Spectrosc.,* **35,** 77 (1981).
43. M. A. Ford, R. A. Spragg, *Appl. Spectr.,* **40,** 715 (1986).
44. R. Messserschmidt, private communication.
45. K. Makinouchi, K. Wagatsuma, W. Suetaka, *Bunkō Kenkyu,* **29,** 23 (1980).
46. (a) M. Primet, P. Fouilloux, B. Imelik, *Surf. Sci.,* **85,** 457 (1979);
     (b) P. C. M. van Woerkom, *J. Mol. Struct.,* **78,** 31 (1982).
47. ref. 45).
48. ref. 42(a).
49. ref. 42(a).

Part II
Selected Applications of
FT-IR Spectroscopy

3.   Analysis of Film and Film-like Samples

3.1   Dichroic Measurements of Drawn Polymer Films

An oriented sample such as a drawn polymer film shows a different absorption or reflection IR spectrum when the orientation of the sample relative to the direction of *linearly polarized light*[*1] is changed, because an interaction between a (polarized) electric field of the radiation and a dipole moment associated with the vibration becomes maximum or minimum when an angle between those two are 0° or 90°.   Thus, dichroic observations of oriented samples allow us to investigate:
(1) the direction of the modes of vibrations,
(2) orientation of molecules or functional groups in the crystalline lattice, and
(3) fraction of the perfect orientation in oriented samples.
Although (1) and (2) are of spectroscopic importance in the study of oriented samples, the last is of industrial interest from the quality control viewpoint, because some physical properties of drawn samples are related to the degree of orientation.   Thus, the dichroic measurement will be used for monitoring of the production of drawn polymers.

This section deals with the dichroic measurements of drawn polymer films.   The incident radiation employed for dichroic observation is the linearly polarized radiation, in which the electric field of the radiation stays in a plane containing the direction of travel of the radiation. In order to capture polarized light, a wire-grid type polarizer or a pile-of-plates polarizer is commonly used.   Polarizers commercially available are:
(1) KRS-5/Al wire-grid type polarizer for $10,000 - 285$ cm^{-1}.[*2]
(2) AgBr/gold wire-grid type polarizer for $5,000 - 285$ cm^{-1}.[*3]
(3) Polyethylene/gold wire-grid type polarizer for $333 - 10$ cm^{-1}.[*2]
(4) Ge "pile-of-plates" polarizer (or Ge Brewster's angle polarizer).[*4]

In the FT-IR system, a AgBr polarizer can deteriorate due to the photo-decomposition of AgBr when unattenuated He-Ne Laser light exists in the sample compartment.

Since the FT-IR spectrometer does not utilize narrow slits or gratings, the residual polarization of the incident radiation is not large.   However, since the residual polarization is *not* zero either, any experimental procedure must be kept free from the difference between the interferometer's transmission efficiencies ($t_{//}$ and $t_{\perp}$) towards parallel and perpendicular polarized light as follows.
(1) Set the polarizer so that the direction of the electric field is 45° away from the vertical line, and scan for the background spectrum.   (2) Mount the drawn polymer film such that the drawn axis is 45° away from the vertical line of the sample holder.   (3) Align the sample mounted in the holder so that the directions of the polarization and the drawn axis are parallel.   (4) Scan for a *parallel absorption* spectrum.   (5) Turn the sample 180° around the vertical line and observe that the direction of the polarization and the drawn axis is now 90° apart or perpendicular.   (6) Scan for a *perpendicular absorption* spectrum.

[*1] Observation using two different polarized radiations whose electric field rotates when the radiation travels is called "circular dichroism; CD" or "optical rotation dispersion; ORD."   These spectroscopies are different from the dichroic observation described in this section.

[*2] Data taken from the Specac product description list.

[*3] Data taken from the Perkin-Elmer product description list.

[*4] 600 cm^{-1} cutoff for Ge polarizer, data taken from the Harrick Scientific brochure.

The absorbance ratio of a pair of parallel and perpendicular absorptions is called the "dichroic ratio" and the dichroic ratio is expressed by

$$R = A_{/\!/}/A = \log(1/T_{/\!/})/\log(1/T_{\perp})$$

Note that the efficiency of the polarizer is not perfect, causing a contribution of the undesired radiation which is 90° away into the observation as stray light.   When the ratio of the stray light to the true polarization is given by t_{rom}, the above-mentioned relationship is modified as

$$R = A_{/\!/}/A_{\perp} = \log[(1 - t_m)/(T_{/\!/} - t_m \cdot T_{\perp})]/\log[(1 - t_m)/(T_{\perp} - t_m \cdot T_{/\!/})]$$

Figure 3.1 shows the t_m spectrum of the AgBr polarizer which the present authors used. Although the t_m value becomes less than 4%T below 2,000 cm^{-1}, the high value of t_m shows that the influence of the imperfect polarizer is not negligible in the spectral region higher than 2,000 cm^{-1}.   The reader should measure the t_m value and use the modified equation for the dichroic ratio determinations.

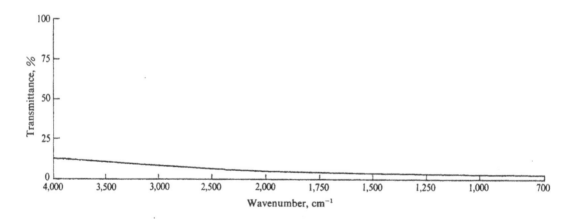

Fig. 3.1   Frequency dependence of t_m of an AgBr wire-grid polarizer.

3.1A	**Presentations of Dichroic Data**	

In this section, four methods of presenting dichroic data are described. By definition, dichroic ratio, R, is given by $R = A(\tilde{v})_{//} / A(\tilde{v})_{\perp}$. The simplest method to show the dichroic features is to show the parallel and perpendicular absorptions in one chart, as seen commonly in the literature. Spectrum (a) exemplifies this method, in which the perpendicular (upper) and the parallel (lower) absorption spectra of drawn poly(propylene) film are shown. It is evident that the absorptions at 1,169, 998, 973, and 841 cm^{-1} show strong perpendicular absorptions, while the absorption at 1,377 cm^{-1} is a strong parallel absorption.

Spectrum (b) shows a difference spectrum, $(A(\tilde{v})_{//} - A(\tilde{v})_{\perp})$. As the above quantity reveals, positive and negative peaks correspond to parallel and perpendicular absorptions, respectively. The difference spectrum is given by

$$(A(\tilde{v})_{//} - A(\tilde{v})_{\perp}) = (\alpha(\tilde{v})_{//} - \alpha(\tilde{v})_{\perp}) \cdot c \cdot l$$

where c, l, and $\alpha(\tilde{v})$ stand for a sample concentration, a cell length (or sample thickenss), and an absorption coefficient at frequency $\tilde{v}$ per unit cell length and unit concentration. Since the difference spectrum depends on the concentration and cell length, both c and l must be known when the balue, $(A(\tilde{v})_{//} - A(\tilde{v})_{\perp})$, is to be used for the determination of the dichroic ratio. A misleading aspect of the difference spectrum is that the amplitude of the difference spectrum, $(A(\tilde{v})_{//} - A(\tilde{v})_{\perp})$, is sometimes irrelevant to the dichroic ratio, $A(\tilde{v})_{//} / A(\tilde{v})_{\perp}$. Therefore, a weak peak in a difference spectrum does not necessarily imply that it is a poor measure of the dichroic ratio.

Spectrum (c) shows a dichroic ratio spectrum, $A(\tilde{v})_{//} / A(\tilde{v})_{\perp}$, resulting from the calculation using the two spectra shown in spectrum (a).

This ratio spectrum is given by

$$R(\tilde{v}) = A(\tilde{v})_{//} / A(\tilde{v})_{\perp} = \alpha(\tilde{v})_{//} \cdot c \cdot l / \alpha(\tilde{v})_{\perp} \cdot c \cdot l = \alpha(\tilde{v})_{//} / \alpha(\tilde{v})_{\perp}$$

This value is independent of concentration and cell length, seemingly showing the dichroic ratio directly. Therefore, this type of display is shown in many publications. The shortcoming of this display is that it is very asymmetric, the parallel polarization ranges from 1 to ∞ while the perpendicular polarization ranges from 0 to 1.

Spectrum (d) shows $R'(\tilde{v}) = (A(\tilde{v})_{//} - A(\tilde{v})_{\perp}) / (A(\tilde{v})_{//} + A(\tilde{v})_{\perp})$, which removes the asymmetry in displaying the dichroic data. In this display, in which the value is independent of concentration and cell length, the parallel polarization ranges from 0 to 1 and the perpendicular polarization from 0 to -1. Because of the symmetrical display of the dichroic ratio, this method also appears in the literatures.

Comparing spectrum (b) with dichroic ratio, (c) or (d), the reader will realize that the peak at 900 cm^{-1} shows a large dichroic ratio, while the difference spectrum (b) shows only a weak indication of a dichroism at 900 cm^{-1}.

It must be noted that spectra (c) and (d) do not give the dichroic ratio data directly, since those spectra are calculated using the computer system of a commercial FT-IR system containing an absorbance offset corresponding to an energy loss due to the reflection at the sample surface in each absorbance value.*

---

* Without including this absorbance, $A_{//}$ and $A_{\perp}$ values away from the band are both 0 absorbance, causing computer error on calculating $A_{//}/A_{\perp}$.

Chart 3.1A

<Experimental conditions>
4 cm^{-1} resolution, DTGS detector, 16 scans coadded, AgBr wire-grid polarizer

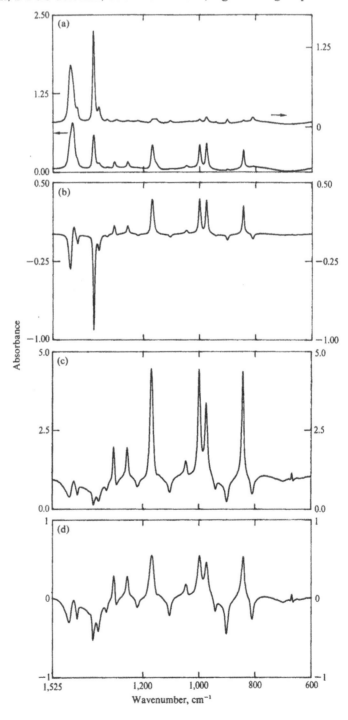

3.1B	Dichroic Spectra of Poly(ethylene)	

In this section, dichroic spectra of drawn poly(ethylene) (PE) film are shown. To analyze IR spectra, PE can be approximated as a long chain of repeating CH_2 units, with the assumption that the contributions from functional groups at both terminals, branching, and double bond formation are negligible. The normal modes of vibration for this model are well established both experimentally and theoretically.[1] Fig. 3.2 illustrates each mode of vibration, a symmetric stretching $\nu_s(CH_2)$ (a), a wagging w_{CH_2} (b), a twisting t_{CH_2} (c), an asymmetric stretching $\nu_a(CH_2)$ (d), a rocking rCH_2 (e), and a scissors δCH_2 (f) vibrations of the CH_2 group. In addition, there are three more modes of vibration which are described in terms of skeletal C–C vibrations.

In the case of an IR active symmetric CH_2 stretching vibration (a), for example, when one CH bonding stretches (shrinks), the other stretches (shrinks) and as a result the induced dipole moment increases (decreases) along the x axis as shown in Fig. 3.2(a). When the direction of the induced dipole moment (called "vibrational transition moment") m coincides with that of the electric field of IR radiation, the absorption intensity is the maximum, for example I. When the angle between those two is 90°, the absorption intensity becomes 0. At angle θ the absorption intensity is given by

$$I_\theta = I \cdot \cos^2 \theta \tag{3.1}$$

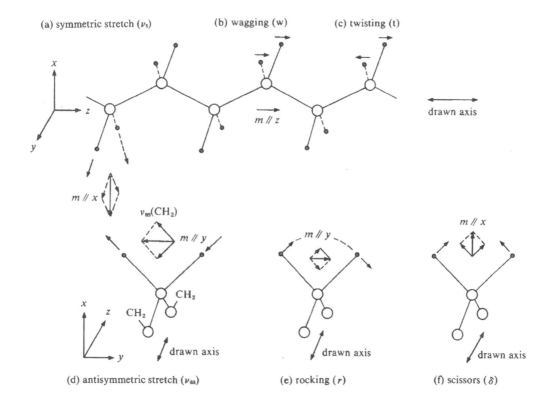

Fig. 3.2 The standard modes of poly(ethylene) vibration.

Since PE molecules align along the drawn axis, the absorption intensity of the $\nu_s(CH_2)$ vibration reaches maximum when the electric field of the polarized IR radiation is normal to the drawn axis.   Thus, $\nu_s(CH_2)$ shows "perpendicular dichroism."

On the other hand, in the case of the CH_2 wagging vibration it is clear from Fig. 3.2(b) that the absorption intensity reaches maximum when the direction of the drawn axis is set to the direction of the electric field, the z axis.   The wCH_2 exhibits a "parallel absorption."

The chart presents the dichroic spectra of the drawn PE film.   Being consistent with the directions of the transition moments shown in Fig. 3.2(a) and Fig. 3.2(d), $\nu_s(CH_2)$ and $\nu_{as}(CH_2)$ show a strong perpendicular dichroism at 2,927 and 2,852 cm^{-1}, respectively.   Two doublet bands, δCH_2 (Fig. 3.2(f)) at 1,475 and 1,463 cm^{-1} and rCH_2 (Fig. 3.2(e)) at 731 and 720 cm^{-1}, also show a perpendicular dichroism as expected from Fig. 3.2.   The aforementioned wagging vibration is assigned to the band with parallel dichroism at 1,176 cm^{-1}.   Although the resultant dipole moment of tCH_2 vibration becomes 0 according to Fig. 3.2(c), which suggests the is IR inactive, this band is assigned to a weak band at 1,050 cm^{-1} with parallel dichroism.

1.   M. Tasumi, T. Shimanouchi, and T. Miyazawa, *J. Mol. Spectrosc.*, **9**, 261(1962); *ibid.*, **11**, 422 (1963).

Chart 3.1B

<Experimental conditions>
4 cm^{-1} resolution, DTGS detector, 10 scans (Ordinates of ⊥ spectra offset)

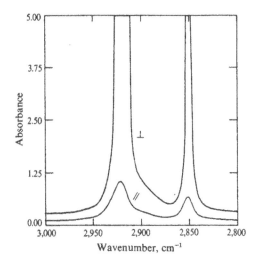

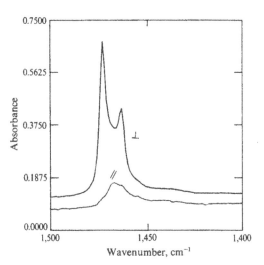

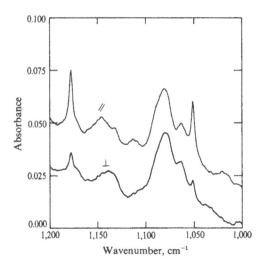

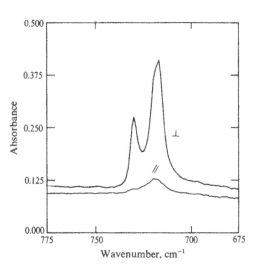

| 3.1C | Orientation of Drawn Poly(propylene) Film (I) | Transmission |

Paper towels wrapped in plastic bags are often used in cafeterias.   The plastic bag we were interested in tore off easily when it was stretched in the direction of the long axis of the bag, while the bag was quite durable to stretching in the short axis.   The material of the bag was easily assigned to an isotactic poly(propylene),[1] PP.   In order to explain the anisotropic strength, we measured the dichroic spectra of this film.   As stated in Section 3.1, the film was mounted in a paper mount with the film's long axis 45° to the long axis of the mount, and two spectra, one of which was to be a parallel absorption spectrum and the other a perpendicular absorption spectrum, were obtained.

In the case of a mono-axially drawn sample such as drawn fiber and drawn plastic film where the polymer chain is oriented in only one direction, the intensities of parallel, $A_{//}$, and perpendicular, $A_\perp$, absorptions are expressed in terms of $A_{//} \propto m(\mu)^2\cos^2\alpha$, $A_\perp \propto (m/2)(\mu)^2\sin^2\alpha$, respectively, where m, μ, and α are the mass of the oscillator, the transition moment, and the angle between the electric field and transition moment.   The dichroic ratio is thus given by

$$R_0 = A_{//}/A_\perp = 2\cot^2\alpha \tag{3.2}$$

When the fraction f is perfectly oriented and fraction $(1-f)$ is randomly oriented, dichroic ratio is given by[2]

$$R = \frac{f\cos^2\alpha + 1/3(1-f)}{1/2 f\sin^2\alpha + 1/3(1-f)} \tag{3.3}$$

Or, rewriting the equation, we obtain

$$f = \left\{\frac{(R-1)}{(R+2)}\right\} / \left\{\frac{(3\cos^2\alpha - 1)}{2}\right\} = \left\{\frac{(R-1)(R_0+2)}{(R+2)(R_0-1)}\right\} \tag{3.4}$$

Therefore, if α is known, the fraction, f, of the oriented portion is obtained from the observed R value. R_0 is the dichroic ratio for the perfect orientation.

The chart shows a parallel* (a) and a perpendicular (b) absorption spectrum of the PP film. Since differences in the intensities of all of the absorption bands are rather small, it is predictable that the degree of the perfect orientation is rather small.

In Table 3.1, parallel and perpendicular intensities of the 15 absorption bands shown in the chart are listed, together with the assignment given by Miyazawa et al.[3]   Using the dichroic ratios of 998 and 841 cm^{-1} bands, $R_{998} = 1.562$ and $R_{841} = 1.596$, in Eq. 3.4, $f_{998} = 0.16$ and $f_{841} = 0.17$ were obtained, provided that the α value of these absorptions was 0°.   From the observed dichroic ratio of the peaks at 899 and 809 cm^{-1}, $R_{899} = 0.755$ and $R_{809} = 0.753$, with the reasonable assumption that the α value was 90°, we obtained $f_{899} = 0.18$ and $f_{809} = 0.18$.   Thus, we conslued the degree of orientation of this poly(propylene) film to be about 17.3% (averaged value).

Since a three-fold helical axis of poly(propylene) molecules is aligned to the drawn axis, the molecules are easily separated, i.e., torn, when the film is stretched in the direction perpendicular to the draw axis.   This force overcomes the van der Waals force between the polymer chains and film tears.   On the other hand, when the film is stretched along the draw axis, the force is weaker

---

* The drawn axis was assigned to the short axis of the bag based on the fact that the intensity ratios of all of the bands are consistent with Table 3.1.

than the C–C bond energy and film is not torn so easily.   Thus all of the dichroic data are consistent with the assumption that the PP film was stretched in the direction of the short axis.

TABLE 3.1   Classification of PP-absorption Intensities

cm^{-1}	A$_{//}$	A$_{\perp}$	R	Assignment	Note
1,460(3)	0.8962	0.9128	0.982	$\delta_a CH_3$	C, A, $\perp$
1,378(8)	0.9774	0.9905	0.987	$\delta_a CH_3$	
1,330(0)	0.1071	0.1216	0.881	wCH_2+CH bending	$\perp$
1,304(4)	0.1900	0.1349	1.408	wCH_2+tCH_2	//
1,245(6)	0.1655	0.1152	1.437	tCH_2+CH bending	C, A, //
1,219(9)	0.0513	0.0605	0.848	tCH_2+rCC	$\perp$
1,168(8)	0.5303	0.3696	1.435	$rCC+rCH_3$	C, //
1,103(3)	0.0747	0.0864	0.865	rCH_3+rCC	C, $\perp$
1,044(4)	0.0611	0.0361	1.693	$rC-CH_3+rCC$	C, //
998(8)	0.5524	0.3536	1.562	$rCH_3+rC-CH_3$	C, helix
973(3)	0.5610	0.4136	1.356	rCH_3+rCC	C, A, //
940(0)	0.0343	0.0417	0.823	rCH_3+rCC	C, $\perp$
899(9)	0.1066	0.1412	0.755	rCH_3+rCH_2	$\perp$
841(1)	0.4903	0.3072	1.596	rCH_3+rCC	C, //
809(9)	0.1082	0.1436	0.753	$rCCH_3+rCH_2$	$\perp$

C: Crystalline band, A: amorphous band, // parallel absorption, $\perp$: perpendicular absorption, δ: bending, w: wagging, t: twisting, r: rocking, rCC: CC stretching, $rC-CH_3$: C–CH$_3$ stretching.

1. Poly(propylene) used for fibers and films are isotactic, although atactic and syndiotactic poly(propylene) are also known.
2. R.D.B. Fraser, *J. Chem. Phys.*, **21**, 1511 (1953); *ibid.*, **28**, 1113 (1958); *ibid.*, **29**, 1428 (1958).
3. (a) T. Miyazawa, *J. Chem. Phys.*, **35**, 693 (1961);
   (b) T. Miyazawa, Y. Ideguchi, K. Fukushima, *J. Chem. Phys.*, **38**, 2709 (1963);
   (c) T. Miyazawa, *J. Polymer Sci.*, **B2**, 847 (1964).

Chart 3.1C

<Experimental conditions>
4 cm^{-1} resolution, DTGS detector, 10 scans.   In order to avoid the appearance of interference fringe patterns, the film was sandwiched between KBr plates.

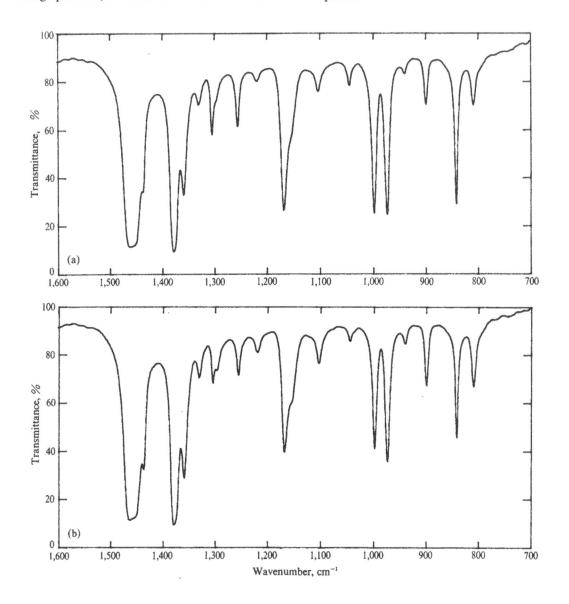

3.1D | Orientation of Drawn Poly(propylene) Film (II) | ATR

The dichroic observation of an oriented sample with the ATR method reported by Flournoy and Schaffers[1] is described in this section. The total reflection of the parallel (↑) and the perpendicular (○) radiations shown in Fig. 3.3 is described. In Fig. 3.3, x, y, and z axes are the drawn axis, normal to the drawn axis of a sample, and the normal axis to the boundary between the IRE and the sample, respectively. The experimental procedure is as follows.

(1) Measure the background spectrum with parallel polarization.[2]
(2) Measure the parallel absorption, $A_{/\!/}{}^1$, by setting the drawn axis along the long axis of the IRE.
(3) Rotate the sample around the z axis to set the principal axis along the y axis and measure $A_{/\!/}{}^2$.[2,3]
(4) Rotate the polarizer 90° and measure the background spectrum for a perpendicular polarization.
(5) Repeat sampling steps (2) and (3) to observe $A_\perp{}^1$ and $A_\perp{}^2$.

The observed values, $A_{/\!/}{}^1$, $A_{/\!/}{}^2$, $A_\perp{}^1$, and $A_\perp{}^2$, are expressed in terms of sample extinction coefficients, k_x, k_y, and k_z, and the experimental parameters described below.[4]

$$2.303A_\perp{}^1 = \beta k_y, \quad 2.303A_{/\!/}{}^1 = \alpha k_x + \gamma k_z,$$

$$2.303A_\perp{}^2 = \beta k_x, \quad 2.303A_{/\!/}{}^2 = \alpha k_x + \gamma k_z$$

where

$$\alpha = \frac{4n_2^2(1 - n_2^2/n_1^2\sin^2\theta)}{n_1^2\tan\theta(1 - n_2^2/n_1^2\sin^2\theta)^{1/2}(1 - n_2^2/n_1^2\sin^2\theta + n_2^4\cos^2\theta/n_1^4)}$$

$$\beta = \frac{4n_2^2}{n_1^2\tan\theta(1 - n_2^2/n_1^2\sin^2\theta)^{1/2}(1 - n_2^2/n_1^2)}$$

and

$$\gamma = \frac{4n_2^2}{n_1^2\tan\theta(1 - n_2^2/n_1^2\sin^2\theta)^{1/2}(1 - n_2^2/n_1^2\sin^2\theta + n_2^4\cos^2/n_1^4)}$$

(3.5)

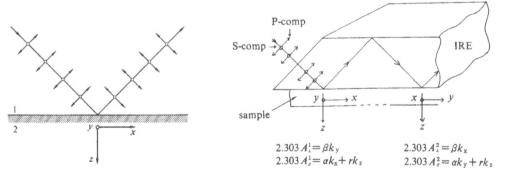

$$2.303\,A_\perp^1 = \beta k_y$$
$$2.303\,A_r^1 = \alpha k_x + r k_z$$

$$2.303\,A_\perp^2 = \beta k_x$$
$$2.303\,A_r^2 = \alpha k_y + r k_z$$

Fig. 3.3 A coordinate system (x, y, and z) and the direction of electric fields in the case of polarized ATR measurements (a) and sample settings for $A_{/\!/,\perp}{}^1$ and $A_{/\!/,\perp}{}^2$ observations (b).

where n_1, n_2, and θ are refractive indices of IRE and sample, and incident angle, respectively. Since only k_x and k_y are determined from the transmission spectra of the film, it is an advantage of the ATR method that all three directions of the transition moments of k_x, k_y, and k_z can be determined for a film.

In order to determine the dichroic ratio of a drawn film, two observations, $A_\perp{}^1$ and $A_\perp{}^2$, using perpendicular polarization suffice.   We examined the sample described in Section 3.1C by the ATR method, in which a KRS-5 prism with a 45° angle of incidence was employed.   Spectra taken from observations with (a) the drawn axis along the long axis and (b) the drawn axis along the short axis of the IRE using s-polarized light are given in the chart.[5]   A peak at 998 cm^{-1}, which shows parallel dichroism, gives $k_x{}^{998} = 0.414$ and $k_y{}^{998} = 0.238$ or $R^{998} = 1.74$.   A peak at 809 cm^{-1} with perpendicular dichroism gives $k_x = 0.110$ and $k_y = 0.146$ or $R^{809} = 0.753$.   A fraction of the perfect orientation is calculated as $f^{998} = 0.20$ and $f^{809} = 0.18$ with Eq. 3.4 given in Section 3.1C.   The dichroic ratio and the fraction of the perfect orientation determined with ATR method compare favorably with the dichroic ratio ($R^{998} = 1.56$, $R^{908} = 0.775$) and the fraction of the perfect orientation ($f^{998} = 0.16$, $f^{908} = 1.18$) determined by the transmission method described in Section 3-1D.

1.   P. A. Flournoy, W. J. Schaffers, *Spectrochim. Act*, **22**, 5 (1966).
2.   The light with electric field parallel or perpendicular to a plane defined by incident and reflected light is called parallel or perpendicular polarized light, respectively.   Do not be confused by the term parallel an perpendicular used in parallel absorption and perpendicular absorption.
3.   This process means the sample is remounted on the IRE.   Because of poor reproducibility of sample-IRE contact, error in the dichroic ratio is inevitable.
4.   We define $\tilde{n} = n + ik$ in this volume.   However, in this section, the relationaship, $\tilde{n} = n(1 + ik)$, as defined by Flournoy and Schaffers is used.
5.   Square the sample to fit the ATR acessory.

Chart 3.1D

<Experimental conditions>
4 cm^{-1} resolution, TGS detector, 16 scans, Wilks type ATR accessory with a KRS-5 IRE at 45°
angle of incidence

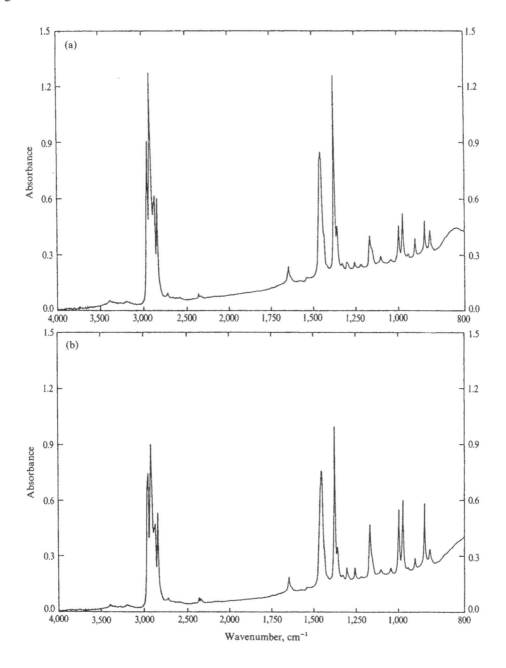

3.2   Analysis of Foreign Matter in Polymers

Polymer products usually have homogeneous surfaces, regardless of whether the surfaces are manufactured smooth or intentionally rough.   However, during production, due to accidental invasion of foreign matter into starting materials, thermal and/or chemical decomposition of starting materials, insufficient mixing or incomplete melting of materials, and so on, the products may have irregular surfaces.   The morphology of irregularity varies.   Bleeding phenomenon occurs when small amounts of some ingredients separate from the mixture and form a thin film on the products.   Fisheye is a typical flaw of transparent polymer films, in which the foreign materials dispersed in the film form small opaque spots.   In some cases the pigment is not mixed well into polymer film, leaving small spots without color.   The most difficult example of irregularity of polymer film is that which can be seen only by polarized light.   In this section some examples of the analysis of foreign matter in polymer products are discussed.

3.2A	Thin Film Formed on Polymer Plate	KBr Pellet (Johnson Method)

A hard black poly(vinyl chloride) plate was brought to our laboratory.   Using reflected light from an incandescent bulb with the grazing angle to the plate, we noted that some parts of the surface reflected the light less because of the bleeding of some material from the surface. Accordingly, the identification of the material on the irregular part of the surface was attempted.

ATR method is known to be one of the best techniques for analyzing the surface.   However, when the sample is not easily mounted in the ATR accessory, the KBr pellet method described below is an excellent method for studying this kind of problem.

Cover the sample surface with 1 to 5 mg of finely ground KBr powder.   Using a flat spatula, rub the surface and powder gently with light pressure.   Make a 1- or 3-mm diameter KBr pellet to obtain a transmission spectrum of the material.

The chart shows the absorption spectrum of the 3-mm diameter KBr pellet made from the poly(vinyl chloride) plate.   A broad intense band centered at 1,426 cm^{-1}, sharp bands at 876 and 712 cm^{-1}, and weak bands at 2,510 and 1,802 cm^{-1} suggest that the main component which has bled to the surface is calcium carbonate, $CaCO_3$, which is the most commonly used filler in poly(vinyl chloride).   In addition to the peaks due to $CaCO_3$, a peak at 1,735 cm^{-1} suggested the presence of some compound with a carbonyl group, most probably an ester.   However, it is not possible to identify the ester using only this single peak.

Although the Johnson method described here is an excellent method for working on this type of problem, the use of diffuse reflection instead of a KBr pellet is also an excellent method, as long as the surface spot is large enough to allow a large enough volume of KBr powder so that a regular (large) sampling cup can be used.

Chart 3.2A

<Experimental conditions>
4 cm^{-1} resolution, TGS detector, 16 scans, 3-mmϕ KBr pellet

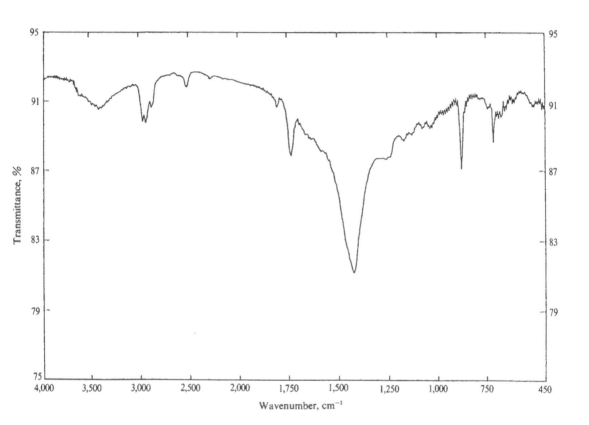

3.2B	Foreign Matter Deposit on Poly(imide) Multilayer Printed Circuit Board	KBr Pellet

The poly(imide) board is a composite product in which thin copper lines are printed on a poly(imide) plate to network an electronic circuit with another poly(imide) plate covering the electric circuit.   Since the poly(imide) multilayer electronic board has flexibility, it is widely used in modern electronic devices as, for examples, stems for thermal printer heads.   It is easy to distinguish poly(imide) from other polymers, since poly(imide) is deep brown in color and transparent.

In the production of the poly(imide) circuit board, many small deposits were found on the surface of the board.   By microscopic observation the size of the deposits were found to range from 50 to 200 μm.   Since the transmission method is the preferred method for obtaining IR spectrum, we decided to make a KBr pellet of this deposit.   First, we placed fine KBr powder on a $\sim$100-μm diameter deposit.   Taking advantage of the fact that poly(imide) does not dissolve in organic solvent, one or two drops of chloroform were poured on the KBr powder using a syringe.   The wet KBr powder was stirred quickly with a spatula.   When the chloroform evaporated, a 1-mmϕ pellet was made and the absorption spectra observed.

The chart presents the IR spectrum obtained from the KBr pellet made by this method.   A sharp absorption at 1,260 cm^{-1}, a broad intense doublet absorption at 1,100 cm^{-1} and 1,000 cm^{-1}, and an absorption at 800 cm^{-1} indicate that the deposit was poly(methyl siloxane), a silicone resin.   Assuming the density of this compound to be around 1 g/ml,[1] a 100 μmϕ hemisphere weighs about 250 ng, which is enough for a good IR spectrum.   Since the KBr powder is chilled as the chloroform evaporates, it absorbs the moisture from the ambient air.   Thus the presence of broad peaks at around 3,400 and 1,640 cm^{-1} is a shortcoming of this method, although the peak at 1,640 cm^{-1} is not evident in the chart.

1. D. O. Hummel, *Infrared Analysis of Polymers, Resins and Additives*, Vol. 1, Part 1, Wiley Interscience, New York (1971).

Chart 3.2B

<Experimental conditions>
4 cm^{-1} resolution, TGS detector, 16 scans

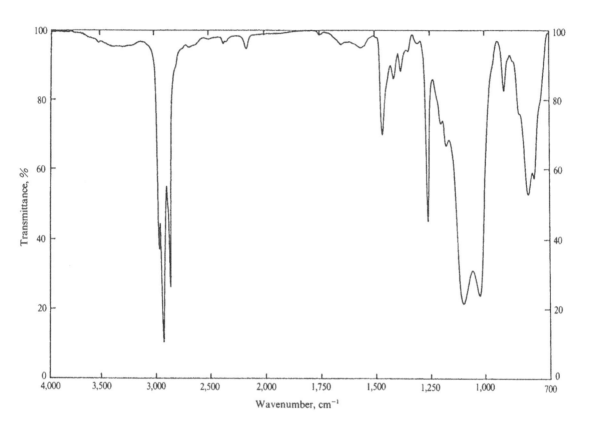

3.2C	Fisheye of Poly(ethylene) Film	Microscope

A sheet of poly(ethylene), PE, mixed with a white pigment, TiO_2, showed that there were a few places where the pigment was not mixed.   Although it was almost impossible to see the flaws with the naked eye, an optical microscope enabled us to identify such places.   The size of the places without pigment was about 20 μm in diameter.   Such a small spot in the sample can be measured only with an IR microscope.

The chart shows the transmission spectra of (a) an irregular spot and (b) a normal spot about 30 μm away from the irregular spot measured using the same microscope and scanning conditions.   An IR spectrum of the irregular spot shows two absorption bands at 1,738 and 1,705 cm^{-1}, both of which are negligible in spectrum (b).   In addition to the two broad absorption bands at 3,300 and 1,650 cm^{-1} as well the hint of another line in the range between 1,500 and 1,600 cm^{-1} seen only in the irregular part.   Another difference exists in the poly(ethylene) spectrum itself.   The intensity of the 1,381 cm^{-1} peak relative to the 1,460 cm^{-1} peak is much stronger in the irregular part than in the normal part.   The sharp cut-off below 800 cm^{-1} in spectrum (b) is due to TiO_2 absorption and the weak absorption in the irregular part agrees well with the nature of the problem, *i.e.* the TiO_2 powder is not well mixed with the poly(ethylene).

A peak at 1,738 cm^{-1} would be assigned to an ester-type carbonyl group, while the 1,705 cm^{-1} peak may be assigned to a ketone-type carbonyl group.   An ester compound is sometimes used as an additive to poly(ethylene).   It is well known that ester, ketone, and acid compounds are formed when poly(ethylene) is oxidized, giving rise to possibility that the ester-type carbonyl compound in question here is an oxidation product.   Bands near 3,300 and 1,650 cm^{-1} with a band near 1,550 cm^{-1} are reasonably assigned to an amide-type absorption.   Long-chain amide compounds such as Erucamide are often added to poly(ethylene) as a slip agent to make poly(ethylene) film slip well on other materials.   However, it is obvious that the concentrations of ester and amide are much higher in the irregular portions of the sample than in the normal portions.   Although the intensity of the 1,381 cm^{-1} peak is used to calculate the content of methyl groups in unit carbon numbers of poly(ethylene), it is known that the intensity of the methyl group is enhanced when a CH_3–C=O group is formed.   Taking the stronger intensity of methyl groups (1,381 cm^{-1}) and the possibility of oxidation (1,705 cm^{-1}) into account, it was assumed that the irregularity was due to the concentration of additives in the oxidized part of poly(ethylene), and these higher concentrations lead to a reduction in TiO_2 content.

Chart 3.2C

<Experimental conditions>
4 cm^{-1} resolution, IR-PLAN microscope with narrow band MCT detector, 500 scans

Fisheye in a Polyethylene Sheet (~20 micron diameter)

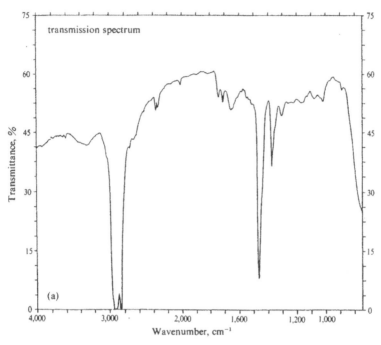

(a)

Normal part of Polyethylene Sheet

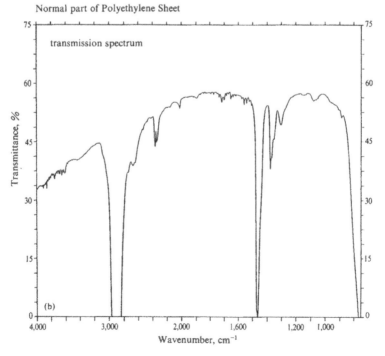

(b)

3.2D	Foreign Matter in Polymer Film (I)	Microscope

Sometimes a variety of tiny foreign matter invades polymer films during production to form tiny irregular spots called "fisheyes," which ruin the value of the products.   Intrusion of foreign matter occurs during the production of polymer itself and/or in the processes of making films. Another cause of fisheye is thermal decomposition or oxidation of the polymer and/or additives during filmmaking due to overheating of the mill and other reasons.   In contrast, when the temperature is too low, some of the materials are not completely melted and result in fisheyes. Usually the size of a fisheye ranges from 10 to a few hundred micromillimeters, requiring the use of an IR microscope for analysis.   We collaborated with the analytical laboratory of one of the poly(propylene), (PP), manufacturing companies to solve the fisheye problem.

An optical microscope image of a fisheye is shown in Fig. 3.4.   The area inside the ellipsoidal white ring looks different from the normal part, and this entire area is thought to be the fisheye when observed by the naked eye.   However, this area merely has a different film thickness or polymer orientation due to the effect of the foreign matter, which is depicted as two white particles in the figure.   Therefore, the target of IR spectra measurement must be those small particles and not the entire area of the fisheye.   Although the long axis of the fisheye was about 600 μm, the particle size was only 50 μm in diameter.

The chart shows the IR spectra of (a) PP obtained from the normal part of the film, (b) the 50-μm particle, and (c) a difference spectrum between (a) and (b) in absorbance units.   As spectrum (a) reveals, the material is isotactic poly(propylene), PP.   Overlapping the PP spectrum, intense broad absorption peaks were observed in spectrum (b) at 3,400 and 1,100 cm^{-1}.   Removing the PP spectrum in difference spectrum (c), we assigned the intense absorption bands to silica or a silicate compound.   This spectrum is most probably that of a sand particle, suggesting that small amounts of sand dust escaped from the air filter system of the factory.

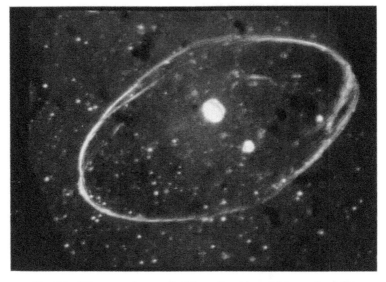

Fig. 3.4   Microscope image of a fisheye found in a poly(propylene) film.

Chart 3.2D

<Experimental conditions>
8 cm^{-1} resolution, IR-PLAN microscope with narrow band MCT detector, 100 scans

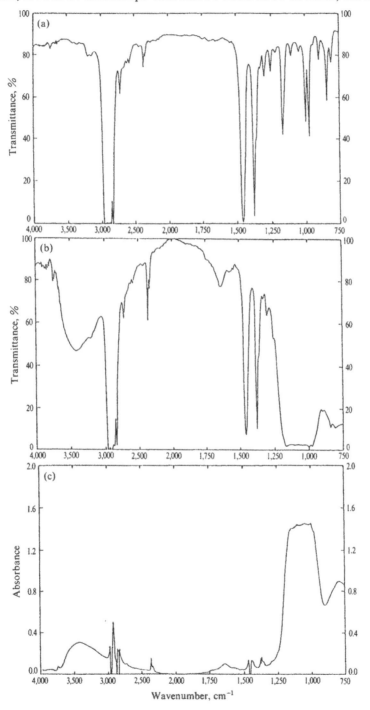

| 3.2E | Foreign Matter in Polymer Film (II) | Microscope |

Additional experiments were performed on the poly(propylene) film described in the preceding section.   In the chart, three examples of particles found in fisheyes are shown.   All are difference spectra in %transmittance units.   Spectrum (a) coincides perfectly with the spectrum of cellulose.   A photograph of the microscopic image of the particle shows that it is a fibrous material, which is consistent with the cellulose identification from the spectrum.   Spectrum (b) is assigned to a polyester such as poly(ethylene terephthalate) or others, based on the presence of a carbonyl group ($1,740 \text{ cm}^{-1}$) and C–O–C ester bands ($1,340$ and $1,100 \text{ cm}^{-1}$) and the pattern of the IR spectrum.   The polymer assignment chart shown in the Appendix was used to confirm that these impurities were cellulose and polyester.   The microscopic image of the particle responsible for this latter spectrum also indicates that it is a fiber.   Spectrum (c) shows a broad band with fine structures at $1,150$, $1,110$, and 970 cm^{-1}.   The features of this broad band compare favorably with IR spectra of $SO_4^=$ salts.   Among sulfuric acid salts which may be used in the factory, $CaSO_4 \cdot 2H_2O$ used as gypsum and $BaSO_4$ used as a paint filler are possible sources. However, a sharp peak at 970 cm^{-1} suggests that the spectrum may be assigned to $BaSO_4$ rather than $CaSO_4 \cdot 2H_2O$.

The peak seen at $2,096 \text{ cm}^{-1}$ in spectrum (c) must be discussed.   Not many kinds of chemical groups give rise to absorption in the frequency range of $2,100 – 2,300 \text{ cm}^{-1}$.   Organic compounds such as nitrile (R–CN), isocyanate (R–N=C=O), isonitrile (R–NC), and azides (R–N=N=N) and inorganic compounds such as CN^- ion, NCS^- ion, NCO^- ion, and their metal complexes show absorption in this region.   Since the absorption frequency coincides with that of a Prussian blue pigment ($Fe_4[(Fe(CN)_6]_3$), we have tentatively assigned this peak to Prussian blue used as pigment in the paint.   This is reasonable since the pigment was found with the paint filler.

Thus, tiny pieces of cellulose fiber, polyester fiber, $BaSO_4$ and Prussian blue were found in the mixture of PP and additives.   They probably fell into the mixture during the production process. The cellulose and polyester fiber may be attributable to a uniform made of cotton and polyester worn by the workers in the factory.   Prussian blue pigment together with paint filler, $BaSO_4$, were also in common use in the factory, in for example walls, ceiling, and or air ducts.

Chart 3.2E

<Experimental conditions>

8 cm^{-1} resolution, IR-PLAN microscope (×32 Cassegrain) with narrow band MCT detector, 50 scans for (a) and (b) and 102 scans for (c)

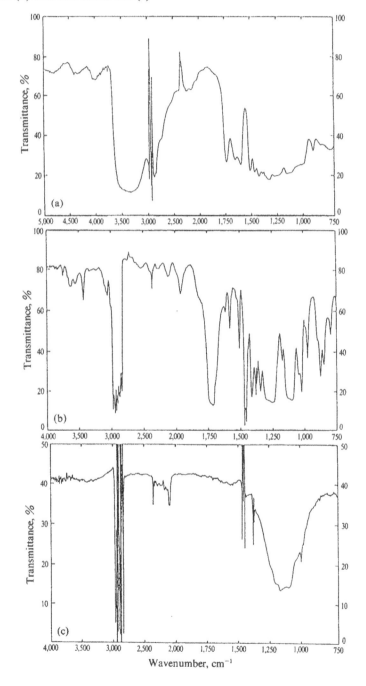

3.3   Layer Structures, Surface Treatments and Chemical Changes in Polymer Films

3.3A	Structure of Eight-Layer Laminate Film	Microscope

Elucidation of the layer structure of a laminate film was attempted.   Laminate film is used in microwavable packaging for storing food in the freezer so the laminate must be durable at low and high temperatures.   A small piece of the laminate film was molded in a methacrylate type resin then sliced into thin film using a microtome.   The thin-sliced sample was placed on a BaF_2 plate and examined by IR microscope.   Optical observation of the cross section showed that the laminate has 8 layers.   The thickness of layers ①, ②, ⑥, ⑦, and ⑧ from the left of Fig. 3.5 is in excess of 80 μm.   Layer ④ is about 60 μm and layers ③ and ⑤ are 10 – 15 μm thick.

The IR spectra of individual layers are shown in the chart.   The spectra of layers ①, ②, ⑥, and ⑦ are identical and easily assigned to poly(propylene).   (Follow the method given in the appendix.   This will lead the investigator to the group that includes poly(propylene).   It can be distinguished from the rest of the polymers by comparing the spectra.)

The spectra of layers ③ and ⑤ are identical and assigned to poly(vinyl acetate) or a copolymer of vinyl acetate and ethylene.   A computer-assisted library search program identifies it as ethylene vinyl acetate copolymer.

Layer ④ showed a typical IR spectrum of poly(vinylidene chloride), which blocks water and gas

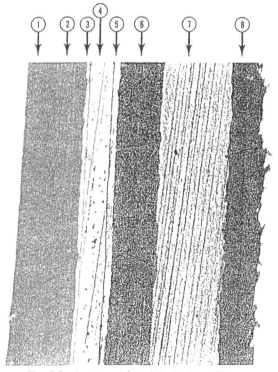

Fig. 3.5   A cross-section of the laminate film.

---

* Data provided by Dr. David Schiering and Mr. Samuel White of Perkin-Elmer Corporation.

Chart 3.3A

<Experimental conditions>
8 cm^{-1} resolution, IR microscope with ×6 Cassegrain and narrow band MCT detector (0.25×0.25 mm element), 16 scans
<Sampling tool>
Microtome: Reichert-Jung 2050 Supercut
Molding resin: Reichert-Jung Historesin

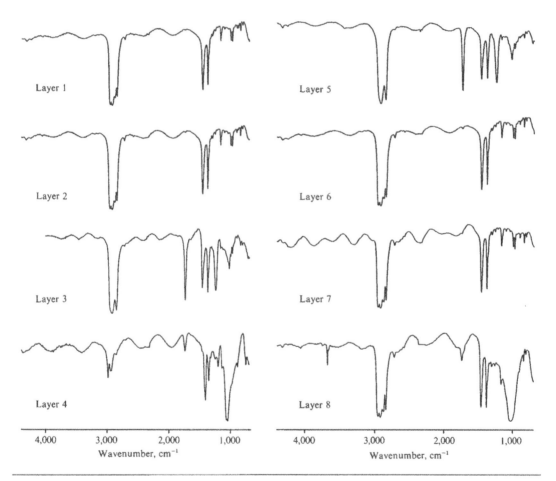

transport through the film.   This layer serves to protect the food from dehydration and prevent oxidation.

Layer ⑧ shows distinctive features of poly(propylene) and two other absorption bands, including a sharp line at 3,700 cm^{-1} and a broad band at 1,100 cm^{-1}.   Indeed, these two lines are indicative of Talc, a magnesium silicate.   Talc is a typical filler used to reduce production cost or reinforce the film.

As shown in this example, the IR microspectroscopy can provide detailed information regarding the structure of laminates.

3.3B	Layer Structure of Automobile Paint	Microscope

Analysis of the layer structure of the paint chip collected at the site of a car accident was performed using an IR microscope.[1]   A microtome was utilized to prepare the section of the paint chip embedded into a methacrylate resin as described in Section 3.3A.   The first layer, the surface, was 100 μm thick and pale blue, the color of the car.   The second layer was 70 μm thick and dark blue.   All remaining layers, the third through eighth layers, were brown and about 60 – 80 μm thick, except for the thinnest fourth layer, which was 40 μm.

The chart shows the microscope transmission spectra of the eight layers.   The first layer is an oil-cured alkyd resin (doublet at 1,600 cm^{-1} characteristic of an ortho substituted benzene ring assigned to a phthalic acid unit; 1,170, 1,160, and 1,070 cm^{-1} ester bands as well as 1,735 cm^{-1}, a carbonyl group).   The second layer showed almost the same IR spectrum with an extra peak at 2,095 cm^{-1}.   It is reasonable to assign this peak to Prussian blue pigment (see Section 3.2.E for assignment), since the color of the second layer is the same color as Prussian blue.   The third and fourth layers are the same compound.   A triplet absorption at 3,700, 3,650 cm^{-1} is characteristic of Kaolinite ([Al$_4$(Si$_4$O$_{10}$)(OH)$_8$], a component of clay) and a doublet absorption at 925 cm^{-1} and an intense broad absorption at 1,100 – 1,000 cm^{-1} support the assignment to silicate.   The rest of the absorptions are attributable to an ester type compound.   Based on the doublet structure of the bands at 1,240 and 1,170 cm^{-1}, the material was determined to be poly(methyl methacrylate). Thus, the third and the fourth layers are an acrylate paint with a clay filler.

The main features of the IR spectra of the fifth layer is similar to that of the first layer, an o-phthalic acid ester.   In addition absorptions characteristic of melamine at 1,550, 1,490, and 815 cm^{-1} were observed.   Thus, the compound was assigned to melamine-cured alkyd resin. The rest of the layers were all determined to be melamine-modified ester type resin because the above-mentioned characteristic absorptions of melamine and ester type resin were observed.

The model and make of the car can be identified referring to a car paint library.[2]

---

[1] Data supplied by Dr. David Schiering of Perkin-Elmer Corporation.
[2] Automotive Paint Library compiled by the Georgia Bureau of Investigation (U.S.A).

Chart 3.3B

<Experimental conditions>
4 cm^{-1} resolution, IR-PLAN microscope with 15× Cassegrain and narrow band MCT (0.25×0.25 mm) detector, 50 scans

<Sampling tool>
Microtome: Reichert-Jung 2050 Supercut
Molding resin: Reichert-Jung Histo Resin

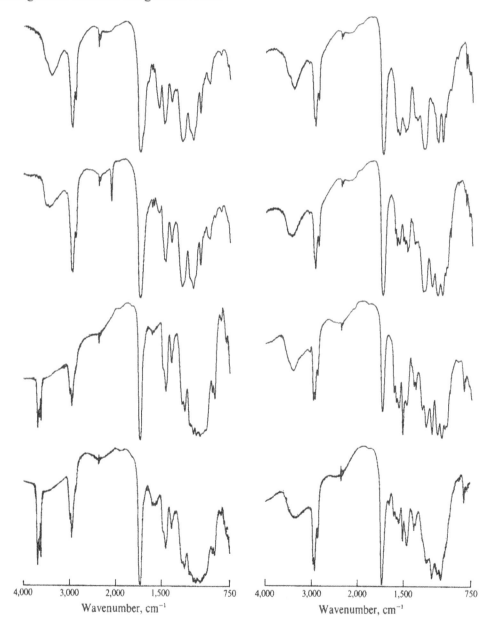

3.3C	**Photochemical Reaction of Polymer (I)**	**Microscope**

Poly(ethylene) undergoes oxidation by ultraviolet light irradiation in ambient air since the photochemical generation of free radicals followed by reaction with oxygen takes place.   As a result, carbonyl compounds are formed at the surface initially and gradually penetrates the interior due to the diffusion of oxygen molecules through the polymer.   In order to monitor the degree of oxidation as a function of depth from surface, a poly(ethylene) sheet subjected to weathering test was examined by IR microscopy.*

The poly(ethylene) sheet was removed from the backup metal block, which prevented oxygen from reacting with free radicals from the backside of the sheet.   The poly(ethylene) was molded in a methacrylate resin and a thin film of the sheet was cut using a microtome to make it possible to perform depth profiling.   The transmission spectrum of an area $12 \times 12 \, \mu m$ was measured every $5 \, \mu m$ from the surface to the backside across the section of the $200 \, \mu m$ thick polymer sheet. As shown in Fig. 3.6, the carbonyl group absorption was observed at $1,735 \text{ cm}^{-1}$ together with a peak at $1,170 \text{ cm}^{-1}$, indicating that the product was an ester type compound.

Stack plot (a) given in the chart shows a decay of intensity of carbonyl compounds as the locus of observation moves from the front surface to the backsurface.   Since the thickness of the sliced film may vary from place to place, peak areas of the carbonyl groups normalized relative to the intensity of the $1,460 \text{ cm}^{-1}$ CH_2 bending vibration were plotted *versus* each depth from the surface in Chart (b).

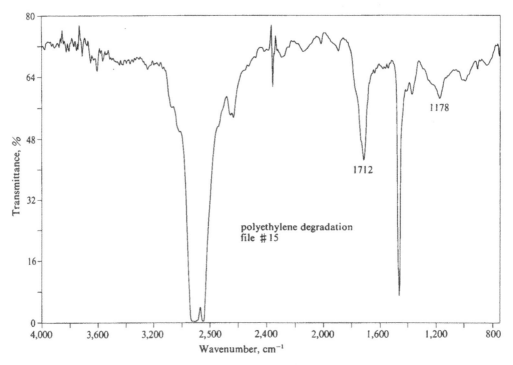

Fig. 3.6   A microscope transmission spectrum of the oxidized poly(ethylene) sheet.

---

* Data provided by Dr. David Schiering of Perkin-Elmer Corporation.

Chart 3.3C

<Experimental conditions>
8 cm^{-1} resolution, IR-PLAN microscope with 15 × Cassegrain, noarow band MCT detector, 128 scans coadded
<Sampling tool>
Microtome: Reichert-Jung Supercut
Molding resin: Reichert-Jung Historesin

(a)

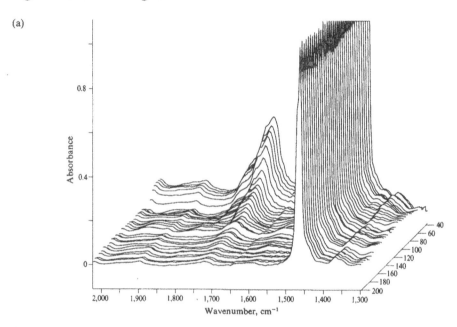

(b)

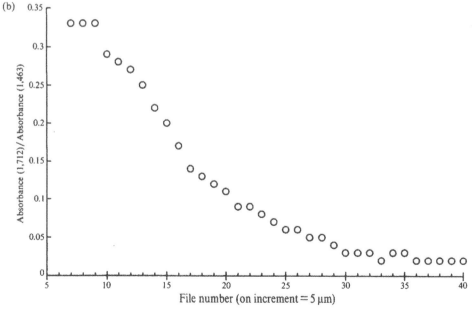

The band shape of the carbonyl group indicates that this band is composed of at least two different bands, suggesting the generation of different kinds of oxidation products by a complex reaction scheme.   This may explain the sudden change in intensity of the carbonyl group at around 55 μm, although the decay seems fairly smooth.

It was shown in this study that IR microspectroscopy can serve as a qualitative and quantitative tool for depth profiling study.

3.3D	**Photochemical Reaction of Polymer (II)**	**PAS**

Since the photochemical oxidation of poly(ethylene) gradually proceeds into the polymer sheet as explained in the preceding section, it is interesting to investigate the position of the reaction frontier.   As explained in Chapter 2, it is possible to set the scanning condition so that the observation thickness, within which the sample contributes to the PA-signal intenstiy, is artificially modified.   The sample described in the previous section was studied by the photoacoustic FT-IR method* and the complimentary results were added to those obtained from the IR microspectroscopy technique.

Let us explain how to set the thermal diffusion length artificially.   Eq. 2.22, as given in Section 2.4,

$$r_t = (2\alpha/\tilde{\omega})^{1/2} = (2k/\tilde{\omega}\rho c)^{1/2} \tag{2.22}$$

requires α or $k/\rho c$ and $\tilde{\omega}$.   Since $\tilde{\omega} = 2\pi f$ and $f = F\tilde{v}$, $\tilde{\omega}$ is easily calculated as the product of the frequency of the interest, $\tilde{v}$ and OPD velocity, F.   If thermal diffusivity is not found easily, look for the thermal conductivity, k, density, ρ, and specific heat, c, from the literature, *e.g.* scientific journals or the CRC Chemical and Physical Constants reference book.

We are interested in the absorption due to the carbonyl group, $\tilde{v} = 1,735\ \mathrm{cm}^{-1}$, in poly(ethylene).   Since the thermal diffusivity of common polymers is about $1.3 \times 10^{-3}\ \mathrm{cm^2\ s^{-1}}$, the thermal diffusion length is calculated to be 6 μm with 0.50 cm s^{-1} OPD velocity and 20 μm with 0.05 cm s^{-1}.

Figure (a) of the chart tells us that the carbonyl compound is already formed 20 μm from the backside of the sheet, since spectrum (a-2) shows the carbonyl band clearly.   On the other hand, spectrum (b-2) in Figure (b) indicates that the carbonyl group is not formed at all 6 μm from the backside of the sheet.

Figure (b) in the chart given in Section 3.3C shows that the carbonyl compound was not detected until the third observation from the back surface, which is 10 – 20 μm from the backside, showing quite good coincidence with the microscopic study.   With the ATR method, another technique allowing depth profiling, one can change the distance which contributes to the spectrum up to *ca.* 1.3 μm for 1,720 cm^{-1}.   Thus PA–FTIR technique has a depth profiling capability of one order of magnitude deeper rom the surface than ATR.

---

* Data provided by Dr. J. McClelland of MTEC Photoacoustic Co.

Chart 3.3D

＜Experimental conditions＞
8 cm⁻¹ resolution, 128 scans. Spectra divided with a background from carbon black. Detector purged with He.

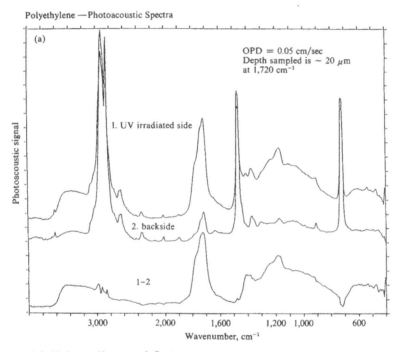

Polyethylene —Photoacoustic Spectra

(a)

OPD = 0.05 cm/sec
Depth sampled is ～ 20 μm
at 1,720 cm⁻¹

Photoacoustic signal

1. UV irradiated side

2. backside

1-2

Wavenumber, cm⁻¹

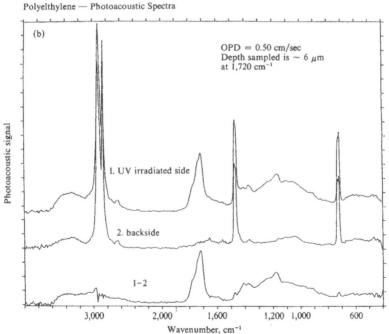

Polyelthylene — Photoacoustic Spectra

(b)

OPD = 0.50 cm/sec
Depth sampled is ～ 6 μm
at 1,720 cm⁻¹

Photoacoustic signal

1. UV irradiated side

2. backside

1-2

Wavenumber, cm⁻¹

3.3E	**Pigment of Glass-reinforced Plastic**[1]	**Microscope**

An experimental method for identifying a pigment in a glass fiber-reinforced poly(phenylene sulfide), PPS, film has been developed.   The resin contains *ca.* 20 weight % quartz fibers. Transmission, ATR, and KBr pellet techniques are available to perform difference spectroscopy to isolate the pigment spectrum from the spectrum containing all of the resin, quartz, and pigment.   However, it is not easy to subtract the reference spectrum taken from material without the pigment from that taken from a sample containing all three components,[2] since the composition of resin and quartz may vary between the sample and the reference material.   In addition, since the quartz has a strong absorption of mid-IR light in the 1,100 – 1,000 cm^{-1} region, the influence of the reflection at the boundary of resin and quartz can appear differently in the sample and the reference.   In other words, a different magnitude of so-called "Restrahlen band" may appear in the sample and the reference spectra, making the difference spectroscopy more difficult.

In order to circumvent this problem, a novel technique was developed using the IR microscope in which the transmission spectra of the sample and the reference films were measured in an area where quartz fibers do not exist.   Thus, the analytical procedure is reduced to perform the difference spectroscopy between the sample with resin and pigment and the reference film composed only of resin, eliminating the problems created by the quartz.   A photograph of a glass fiber reinforced PPS resin is shown in Fig. 3.7 where many quartz fibers are seen to be distributed randomly.   The largest area found was about 30 – 40 × 30 – 40 μm.

Fig. 3.7   A microscope picture of a glass fiber-reinforced PPS sheet.   A bar indicator corresponds to 100 μm.

---

[1] Data provided by Mr. Eiji Hosoda of Tokyo Ink Co. Ltd.
[2] Refer to Section 4A and 4B, for the difference spectroscopy when the sample contains quartz.

Chart 3.3E

<Experimental conditions>
8 cm⁻¹ resolution, 16 scans, IR–PLAN microscope equipped with 32× Cassegrain and narrow band MCT ($0.25 \times 0.25\,\mu$m element)

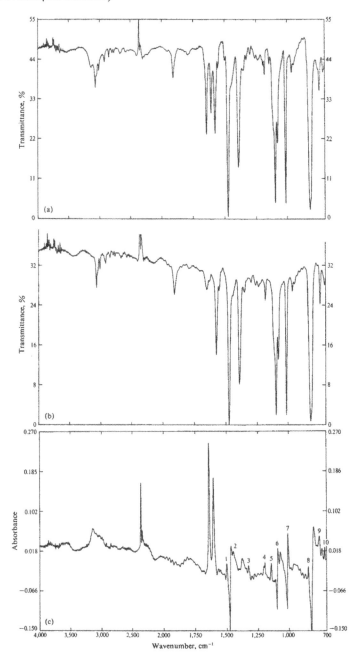

Spectra (a) and (b) in the chart show the transmission spectra of the sample and the reference, respectively.   Neither shows any sign of quartz absorption.   Difference spectrum (c) in the chart clearly shows broad absorptions due to NH or OH at 3,400 cm^{-1}, CH stretching vibrations at 3,150 – 3,000 cm^{-1}, intense doublet band at 1,643 and 1,607 cm^{-1}, and 10 other absorption bands marked with numbers.   All of those peaks and spectral features agree well with those of the spectrum of the diketo-pirolopirole type pigment shown in Fig. 3.8, although derivative type signals due to the incompensation of four strong absorption bands assigned to the PPS resin overlap.

It was found that the microscope is useful to obtain reliable difference spectra from heterogeneous samples in which strong absorbers such as quartz and other inorganic compounds are dispersed (Fig. 3.9).

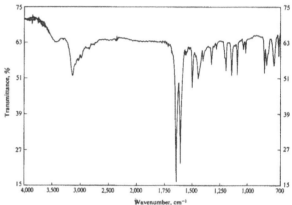

Fig. 3.8   Molecular structure and IR spectrum of the pigment.

Fig. 3.9   Transmission spectra of the sample ▶ and the reference.

| 3.3F | Obtaining IR Spectra of Surface Coating | ATR |

In Section 2.2.4D, how to perform difference spectroscopy of bilayer samples with ATR technique was explained.

This section discusses how the use of Eq. 2.16 in the difference spectroscopy between an ATR spectrum of poly(propylene) sheet coated with poly(styrene) and that of reference poly(propylene) improves the difference spectrum.   A poly(propylene) sheet was covered by a chloroform solution of poly(styrene) and the solvent gradually evaporated to form a thin film of poly(styrene) coating the poly(propylene).

Spectrum (a) of the chart is an ATR spectrum of a polystyrene sheet as a reference.   Since intensities of the peaks at 760 and 700 cm^{-1} are very strong due to the large penetration depth of the evanescent wave in this frequency region, the ATR spectrum was displayed corrected for a constant penetration depth for the entire frequency range as well as the following difference spectra, (b)–(d).

Difference spectrum (b) of the chart was obtained with a macro-language software to compensate the frequency dependence of the penetration depth as given by

$$A_s = A - A_b c_1 \exp(-c_2 \tilde{\nu}) \qquad (2.17)$$

where A_s is the difference spectrum and c_1, c_2, and $\tilde{\nu}$ are experimental constants and the infrared frequency, as explained earlier in Section 2.2.

Spectrum (c) of the chart indicates another example of the difference spectrum using a frequency-independent factor to subtract poly(propylene) spectrum, so that the scaling factor was

Chart 3.3F

<Experimental conditions>
4 cm^{-1} resolution, 36 scans, KRS-5 IRE with 45° of incidence

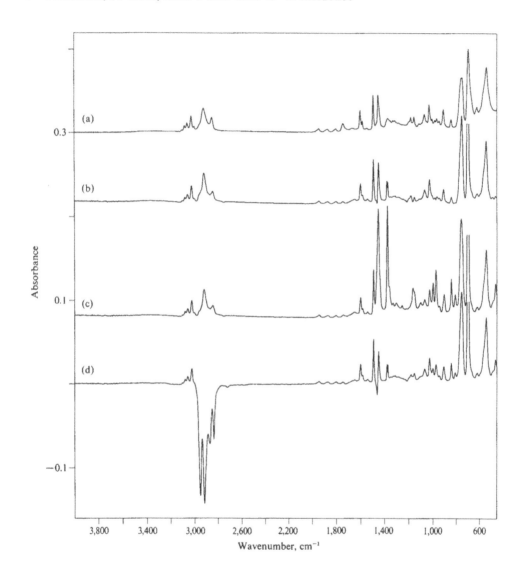

chosen to give a satisfactory difference spectrum in the CH stretching vibration region (3,100 – 2,700 cm^{-1}).   Although overlapping poly(propylene) features in the CH stretching vibrations are removed from the difference spectrum, peaks due to the poly(propylene) are not removed in the frequency region below 2,800 cm^{-1}.   On the other hand, spectrum (d) of the chart was calculated to compensate for a poly(propylene) peak at 1,170 cm^{-1}, using a frequency independent subtraction factor.   As explained in Section 2.2.4D, a subtraction in the region where the frequency is higher than 1,170 cm^{-1} causes over-subtraction as clearly seen in the negative intensities of the CH stretching vibration region (3,000 – 2,700 cm^{-1}).   On the other hand, in the region where the frequency is lower than 1,170 cm^{-1}, under-subtraction of the poly(propylene) spectrum is observed remaining with uncompensated peaks of the poly(propylene).

Although spectrum (b) is largely improved to show a good IR spectrum of the surface compound, poly(styrene), a negative peak at 2,730 cm^{-1} and an incompensation at 1,380 cm^{-1} is obvious.   In addition, strictly speaking, intensity ratios of poly(styrene) peaks thus obtained are not the same as those of the reference spectrum (a).

Thus, as noted in Section 2.2, a perfect compensation of the substrate bands in ATR difference spectroscopy is quite difficult.

3.4   Analysis of Video Tapes, Tires, and Black Samples

3.4A	IR Spectra of Tires (I)	Transmission

Tires are made of vulcanized rubber mixed with carbon powder and some additives.   Natural rubber (NR), ethylene–propylene–diene-monomer rubber (EPDM), styrene–butadiene rubber (SBR) and others are used to manufacture tires.   Since the content of the carbon powder is high, being *ca.* 50 – 60 g per 100 g of rubber, and the carbon is a strong IR light absorber, measurement of the IR transmission spectrum requires the thickness of the tire to be small.   It is also difficult to make KBr pellets from tire samples because grinding a tire chip is difficult due to the elasticity of the tire sample at room temperature.

Spectrum (a) in the chart is the transmission spectrum of a 5×7 mm tire sample, which was cut to 2 μm thick with a microtome and placed on a KBr plate.*   Since the diameter of the IR beam at the sample position is 8 mm, an aperture was made from aluminum foil to mask the part of KBr plate without sample film, so that the IR radiation which did not interact with the sample was not detected.   Since the sample is quite thin, *i.e.* 2 μm, the intensity of the absorption bands due even to the major component, rubber, is quite weak.   In addition, these weak absorption bands overlap on the steep slope of the wide and strong absorption spectrum of carbon black. However, the spectrum is of sufficient quality to identify the rubber as SBR.

Spectrum (b) of the chart is obtained through treatment with a baseline correction routine for sloped baseline.   Several peaks pertinent to each component of the tire are seen.   For example, absorption due to unsaturated CH stretching vibration at 3,100 cm^{-1}, aromatic ring vibrations at 1,600, 1,500, and 700 cm^{-1}, and strong bands due to the *trans*-vinyl group at 963 cm^{-1} and a vinyl group band at 910 cm^{-1} confirm the assignment of the rubber to SBR.

As shown above, slicing the sample by microtome is an excellent method for carbon-rich elastic samples like tires.   However, since thin tire film is apt to oxidized, the measurement should be done immediately after the sample is sliced.

---

* Sample provided by Mr. Kazuhiro Yamada of Yokohama Tire and Rubber Co.

Chart 3.4A

<Experimental conditions>
2 cm^{-1} resolution, DTGS detector, 64 scans

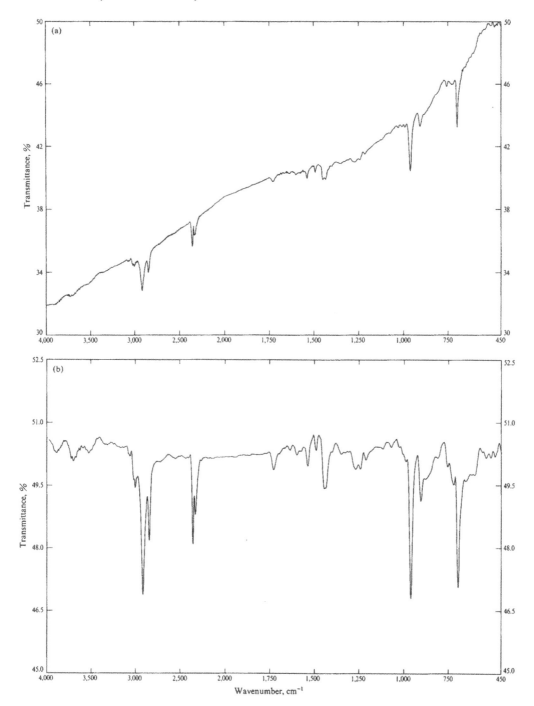

3.4B	IR Spectra of Tires (II)	ATR

We measured the ATR spectra of a series of 11 model tires in which the 0 to 100 g carbon in 10 g increments is added to a mixture of 100 g rubber and additives.   The ATR spectra of samples containing a high level of carbon powder, using KRS-5 or ZnSe IRE with 45° or 60° angle of incidence, will show distortion due to a high refractive index of the sample as explained in Section 2.2.   Therefore, we selected a Ge IRE and 45° incident angle for this work.   Since the optical throughput of the ATR accessory is *ca.* 20% at most, instead of large numbers of repeated scans with a standard OPD velocity such as 0.2 cm/sec, a slower OPD velocity (0.1 cm s^{-1}) with fewer repeated scans was employed (*cf.* Section 1.2.(2))

Spectra (a), (b), and (c) in the chart are observed from model tires[1] with 0, 30, and 60 PHR[2] of carbon powder, respectively.   As the content of the carbon increases, the slope of the spectral baseline increases and the peak intensity of the rubber becomes weaker.   Note that the direction of the slope is opposite to that in the case of the transmission spectrum shown in Section 3.4A. Although the absorption coefficient of the carbon is larger at the high frequency end (4,000 cm^{-1}) of IR spectra than at the low frequency end (700 cm^{-1}) as seen in the transmission spectrum (Fig. 3.6, Section 3.4A), an increase in the absorption intensity of the carbon due to the increase in the penetration depth of the E-wave at 700 cm^{-1} overcomes the decrease in the absorption coeffecient at 700 cm^{-1}, explaining the opposite direction of the solpe.

The relationship between the slope of the ATR spectra and the content of carbon was examined.   In order to correct the difference in the contact between the IRE and the sample, the slope of the baseline (absorbance difference; $A^{702} - A^{4000}$) was normalized using the CH$_2$

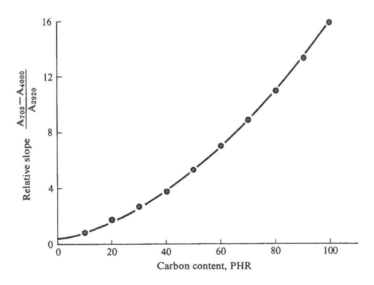

Fig. 3.9   Relationship between relative slope and carbon content.

Chart 3.4B

<Experimental conditions>
4 cm^{-1} resolution, DTGS detector, 36 scans
ATR: Ge IRE 45° angle of incidence

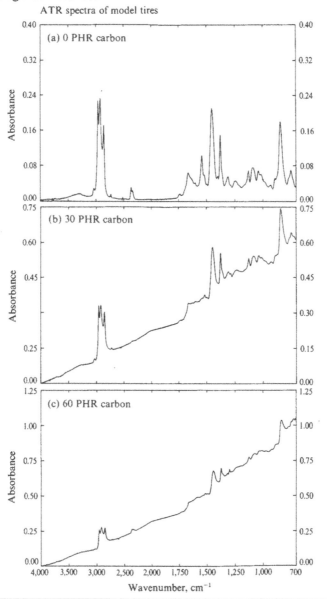

ATR spectra of model tires

stretching vibration at 2,920 cm^{-1}.   The relative slope was plotted *versus* content of carbon in Fig. 3.9.   Since the relationship gives a smooth curve, the carbon content of the tire can be estimated from Fig. 3.9 for tires made of natural rubber.   Needless to say, a similar relationship will be obtained for tires made of different elastomers.

3.4C	Layer Structure of Video Tape (I)	ATR

In a simplified model, video tape is a poly(ethylene terephthalate) (PET) film on whose surface a magnetic material such as γ-ferrite is pasted with an adhesive (called a binder) such as polyurethane.   As such video tape has a distinct multilayer structure.   Although both surfaces of the video tape look black, the nonmagnetic side of most standard grade tapes is colorless and transparent, because the material is untreated PET film.   In such cases, ATR spectra of nonmagnetic and magnetic surfaces are due to PET and magnetic material mixed with adhesive, respectively.   In this section, ATR analysis of the layer structure of high grade video tape, both surfaces of which are coated with black material, is attempted.

As explained by Fig. 2.11 Section 2.2, use of KRS-5 IRE will create severe distortion on ATR spectra of video tape.   Therefore, a Ge IRE with a 45° angle of incidence must be used.

Spectra (a) and (b) in the chart are taken from a magnetic and a non-magnetic layer.   Three absorption bands at 1,720, 1,647, and 1,533 cm^{-1} together with an absorption at 3,300 cm^{-1} (although it is extremely weak) lead us to conclude that the spectrum (a) can be assigned to polyurethane.   Spectrum (b) is also identified as a polyurethane since it is essentially the same spectrum as (a) except for the large baseline slope.   This suggests that the non-magnetic side of the video tape is coated with polyurethane and some material whose ATR spectrum has a wide line-width covering the entire spectral region and a strong absorption.   In addition, the frequency region which is marked by an arrow and is slightly higher than 1,720 cm^{-1} peak, shows a Christiansen-effect type distortion (compare with the same region of spectrum (a)), suggesting that the distortion is caused by the presence of a high refractive index compound such as carbon. The feature of the slope is the same as that observed in the ATR spectra of carbon filled rubber discussed in the preceding section.   Thus, we speculate that the non-magnetic side of the video tape is coated with carbon using polyurethane as a binder.   When Ge is used as an IRE with a 45° angle of incidence, the penetration depth of the evanescent wave at 1,000 cm^{-1} is *ca.* 0.7 μm, as shown in Table 2.1.   Although the evanescent wave penetrates deeper than the penetration depth, the spectrum of the base material, PET, was not observed in either spectrum (a) or (b).   In order to verify that PET is actually used under the surface coatings in this video tape, it is necessary to rely on a photoacoustic (PA) FT-IR technique, since this technique allows penetration one order of magnitude larger than the ATR method.   Contributions to the spectrum from deeper within the sample as measured by the PA-FTIR method is discussed in the following section.

Chart 3.4C

<Experimental conditions>
4 cm^{-1} resolution, DTGS detector, 25 scans
ATR conditions: Ge, 45° angle of incidence

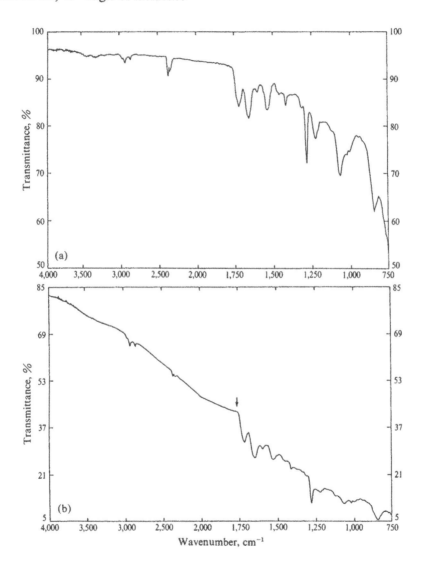

3.4D	Layer Structure of Video Tape (II)	PAS

In the ATR study of the video tape shown in the preceding section, surface thicknesses contributing to the ATR spectra were ~0.5, ~0.7, and ~0.9 μm for 1,700, 1,000, and 750 cm^{-1} of IR light, respectively.   The contribution from locations deeper than the ATR penetration depth must be investigated by the PA–FTIR method, as discussed in Section 2.3.   The video tape examined in the preceding section was tested by the PA–FTIR technique to investigate the structure of layers deeper than those that can be reached by the ATR method.

The two-sided tape was placed on the nonmagnetic side and cut into a circular piece with a paper punch.   This piece was pasted firmly to the sample cup of the PA detector so that there was no air gap between the tape and cup.[*1]

Single beam PA–FTIR spectra (a), (b), and (c) in the chart illustrate how intensities of bands relevant to PET, binder, and magnetic material change with OPD velocity of 0.10, 0.75, and 1.75 cm s^{-1}, respectively.   Since the shallower sample surface contributes to PA–FTIR spectra with faster OPD velocity, the peak intensities due to the magnetic material and binder should become more significant compared with those of the base film, PET.   One of the most prominent bands, a band at 1,720 cm^{-1} of polyurethane overlaps with that of PET at 1,740 cm^{-1}, although the other at 1,650 cm^{-1} is identified in all spectra.   A broad band at 690 cm^{-1}, which is assigned to γ-Fe$_2$O$_3$ based on coincidence with the standard spectrum, becomes stronger as the OPD velocity becomes faster compared to an adjacent peak at 740 cm^{-1} due both to the binder and PET.

Since the thermal diffusivity is calculated as $\alpha = 0.3 \times 10^{-3}$ [*2] and the frequency of interest is 690 cm^{-1}, Eq. 2.23 indicates that the thermal diffusion length varies from 3 to 17 μm when the OPD velocity was altered from 6 to 0.05 cm s^{-1}.   In Fig. 3.10 the intensities of the 690 cm^{-1} peak

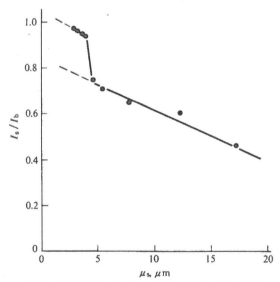

Fig. 3.10   Photoacoustic signal intensity due to γ-ferrite as a function of thermal diffusion length.

[*1]   To avoid interference between photoacoustic signals from the front and backside surfaces.   See N. Teramae and S. Tanaka, *Anal. Chem.*, **57**, 95 (1985).

[*2]   Specific heat at 400°K, $c = 1.32$ cal K g^{-1} and room temperature density $\rho = 5.24$, and thermal conductivity $k = 21.0 \times 10^{-4}$ (cal cm)/(cm^2 s deg) of γ-Fe$_2$O$_3$ were used to calculate $\alpha = k/\rho c$.

Chart 3.4D

<Experimental conditions>
8 cm^{-1} resolution, DTGS detector, MTEC PAS detector model 200, sample chamber purged with
He

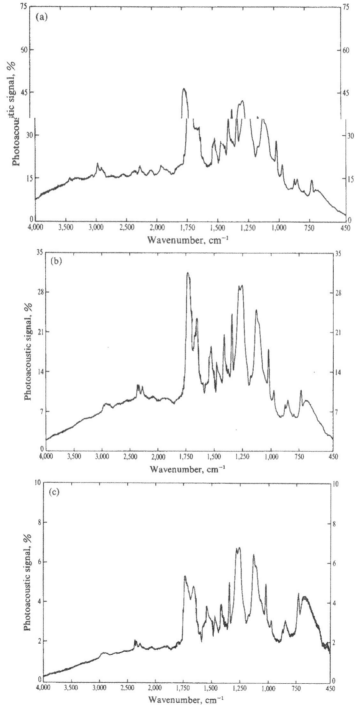

relative to that of $740 \, \text{cm}^{-1}$ were plotted versus thermal diffusion length.   As the thermal diffustion length decreases, the relative intensity of the 690 band due to $\gamma\text{-Fe}_2\text{O}_3$ increases linearly.   However, at the thermal diffusion length of $4.0 \, \mu\text{m}$ the ordinate value jumps to another linear line.   It is reasonable to attribute this jump to the bilayer structure of the video tape.   Thus the thickness of the magnetic layer was concluded to be $\sim 4 \, \mu\text{m}$.

In addition to the PA–FTIR capability of depth profiling, the PA-FTIR method has one more advantage over the ATR method.   Since Ge IRE restricts the lower frequency limit to *ca.* $750 \, \text{cm}^{-1}$, it is not possible to observe the spectrum of the magnetic material.   However, as demonstrated in this section, the PA–FTIR method makes the detection of magnetic material possible.

3.5   Analysis of Printed Substances

3.5A	Peeling of Prints (I)	ATR

A polymer box used for packaging is printed with the name of its product and manufacturer and other information.   The polymer container we analyzed was made of PET, with printing on its surface.   In order to protect the print it was coated with poly(ethylene) (PE) using a two-mix type polyurethane as an adhesive.   Thus the box had a 4-layer structure: PET, printing ink, polyurethane, and PE.   Peeling occurred at the boundary of the ink and polyurethane.

We were asked to investigate the peeling.   An ATR technique was employed, since the analysis involved the surface as opposed to the body of the box.   Spectrum (a) in the chart shows the $4,000 - 2,000 \, \text{cm}^{-1}$ region of the ATR spectra taken from the polyurethane side of the PE surface peeled from the ink.   Spectrum (b) is taken from the PE sheet coated with polyurethane, the coating having been performed using the same materials and reaction conditions as in the manufacturing procedure.

It is evident that the defective polyurethane layer does not show an absorption at $2,280 \, \text{cm}^{-1}$, which is assigned to a residual isocyanate group of one of the two starting reagents that form polyurehane.   In addition, an OH group absorption at around $3,600 - 3,400 \, \text{cm}^{-1}$ is seen in spectrum (a).

$$n \, \text{OCN–R–NCO} + n \, \text{HO–R'–OH} \longrightarrow (\text{OCHN–}R\text{–NHCO–}R'\text{–O})_n -$$

di-isocyanate   diol   polyurethane

Since a slight excess of the isocyanate compound is used in the manufacturing, a residual isocyanate band is usually observed in normally polymerized polyurethane, as shown in Fig. (b). Since the isocyanate absorption is not seen in the polyurethane here and an extra OH group was found instead, it was thought that the reaction to form urethane bonding had not been completed. We concluded that the isocyanate compound was already deactivated when it was used to glue the PE sheet.   Since isocyanate is usually quite reactive, it was thought that the isocyanate had lost its activity due to some kind of reaction, most probably with the moisture in the air during storage.

Chart 3.5A

<Experimental conditions>
4 cm^{-1} resolution, DTGS detector, 16 scans
ATR conditions: Ge IRE, 45° angle of incidence

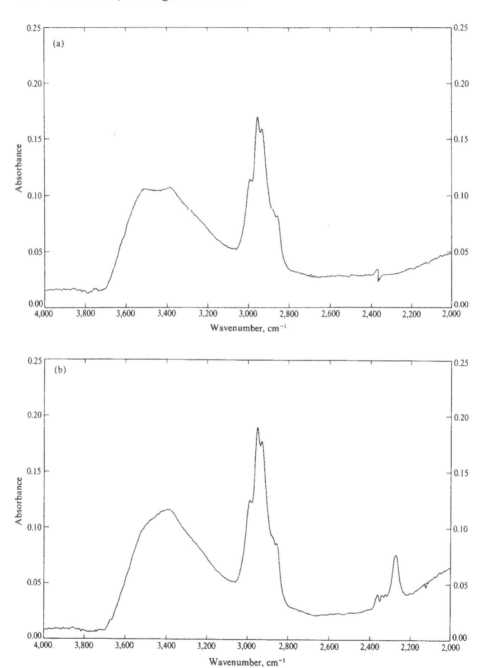

3.5B	Peeling of Prints (II)	ATR

A peeling at the printing interface similar to the preceding example was reported.   However, the peeling in this case occurred at the boundary between the polyurethane and PE sheet.   In addition, an absorption band due to the isocyanate group was found in the ATR spectrum of the polyurethane layer.   Therefore, the cause of this problem was believed to be due to the poly(ethylene) (PE) sheet and not to polyurethane.   As in the preceding section, ATR technique was employed.

Spectra in the chart are those observed from PE sheets; (a) exhibited a peeling problem while (b) did not.   A difference between spectra (a) and (b) shows very strong carbonyl absorption bands at $1,740 \, cm^{-1}$.   The PE sheet used for this particular product contains an ester type compound of aliphatic acid as an additive to prevent water molecules from transmigrating through the sheet.   Since the PE sheet was tested by the manufacturer and found to contain the appropriate amounts of additives, it is reasonable to speculate that the additive concentrated at the surface of the sheet due probably to problems during production.

In order to quantitify this phenomenon, it is necessary to compare the concentration of carbonyl groups at the surface.   Since the intensity of the ATR spectrum is proportional to the degree of contact between the sample and IRE and the evaluation of the contact is difficult, the use of a certain internal standard is generally preferred.   The disparity of sample-to-sample degree of contact is compensated by ratioing the carbonyl absorbance to the internal standard absorbance.   It is convenient to choose the band at $1,460 \, cm^{-1}$ as an internal standard, since the band has enough intensity and the frequency is close to the carbonyl band.   Thus, the value, A^{1740}/A^{1460}, can be used for the receiving test.

Chart 3.5B

<Experimental conditions>
4 cm^{-1} resolution, DTGS detector, 25 scans
ATR conditions: Ge IRE, 45° angle of incidence

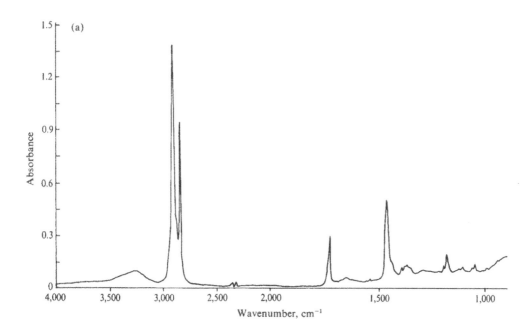

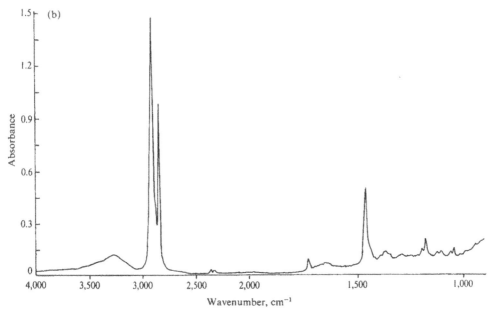

4. Analysis of Fibers and Clothes

Fibers have spherical sections in most cases.   In some cases, the surface may be treated and have a double layer structure as shown in Fig. 4.1.   Transmission spectra of fibers contain light components such as (1) stray light, (2) radiation reflected by the surface, and (3) radiation transmitted through the fiber with different pass lengths, as shown in Fig. 4.1.   Therefore, strictly speaking, the Lambert-Beer law will not hold for fiber and woven cloth samples, their spectra being different from common transmission spectra measured from samples of uniform thickness and no stray light.   However, since it is assumed that the Lambert-Beer law holds for those peaks with weak intensity[1], it may be possible in such cases to obtain a reasonable difference spectrum, allowing identification of the surface treatment material and/or its orientation etc.   Examples of fiber and cloth analyses are given in this section.

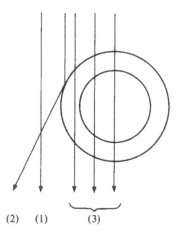

(2)   (1)   (3)

Fig. 4.1   Cross section of a model fiber.   Rays, (1), (2), and (3), represent stray light, reflected light, and light passing through the fiber, respectively.

1.   A. Hanning, *Appl. Spectrosc.*, **42**, 90 (1988).

4A	Observation of Silane Coupling Reagent Bonded to Surface of Glass Fiber (I)	Transmission

A cloth woven with glass fiber, called glass cloth hereafter, is usually used to reinforce plastic board such as that used for printed circuit boards, and the surface of the glass fiber is treated with reagents called "silane coupling reagent" in order to increase its affinity to plastics.   We have tried to isolate the spectrum of the silane coupling reagent from glass cloth using the transmission method.[1]   The glass cloth was placed in the sample position like a polymer film.   Spectra in the chart are observed from (a) surface treated glass cloth, (b) untreated reference glass cloth, (c) reference glass cloth treated with the same procedure as (a) without the silane coupling reagent. Spectrum (d) is the difference spectrum between (a) and (c), and spectrum (e) is the spectrum of γ-anilino propyl tri-methoxy silane (AnPS), the silane coupling reagent used to treat the glass cloth.   Spectrum (c) shows a close resemblance with the diffuse reflection spectrum of neat silica powder.[2]   Since the method used, the transmission method, contains both reflected and refracted radiation, such a resemblance is expected.   Spectrum (a) shows some weak absorption bands due to the treatment, although the major features are the same as those of spectrum (c). The difference spectrum shows fairly good coincidence with the reference spectrum, showing that the silane coupling reagent maintains some chemical structure of the functional groups.   For instance, sharp intense bands at 1,610 and 1,500 cm^{-1} of the reference AnPS are confirmed in the difference spectrum.   However, the CH stretching vibrations in the $3,100 - 2,800$ cm^{-1} region do not coincide well with those of the reference spectrum, suggesting change in chemical structure due to chemical bonding between silica and silane coupling reagent.   Some line widths are also larger than in the reference spectrum.   The most striking discrepancy is in the 1,100 cm^{-1} region where an intense Si–O– band should be observed as in the reference spectrum.   However, since spectral subtraction is performd between the spectra, both of which have an intense Si–O– band of silica, it is difficult to obtain meaningful difference values across this region.   It should be noted that any attempt to calculate a difference spectrum between treated sample (a) and untreated sample (b) results in failure, as the difference in the spectral patterns of sample (a) and (b) is revealed in the 1,500 to 1,000 cm^{-1} region.

1.   N. Ikuta, T. Sakamoto, T. Kouyama, I. Abe, T. Hirashima, *Sen-i Gakkaishi (Fiber Journal)*, **43**, 313 (1987) (in Japanese).
2.   H. Maulhardt and D. Kunath, *Appl. Spectrosc.*, **34**, 383 (1980).

Chart 4A

<Experimental conditions>
4 cm^{-1} resolution, DTGS detector, 36 scans

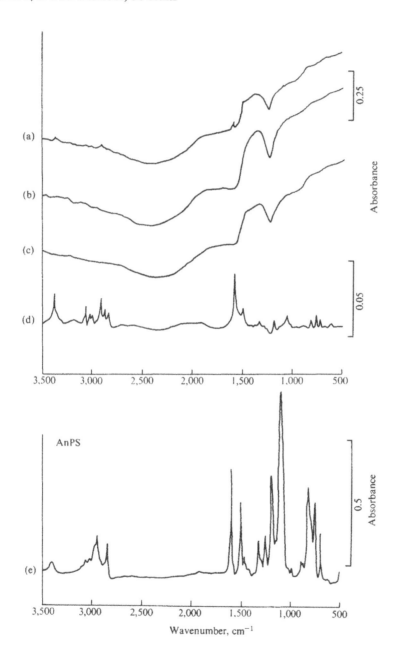

4B	Observation of Silane Coupling Reagent Bonded to Surface of Glass Fiber (II)	KBr Pellet

A KBr pellet method was tried[1] to isolate the spectrum due to silane coupling reagent bonded to the surface of glass fiber, which was discussed in the preceding section. Instead of measuring the glass cloth directly by transmission method, KBr pellets technique was attempted. Using a mortar and pestle, a small amount of the glass cloth was ground into fine powder. A KBr pellet of the glass cloth was thus prepared. Spectra (a) and (b) in the chart were taken from glass cloth treated with and without silane coupling reagent, respectively. Although spectrum (a) looks exactly the same as spectrum (b), an extremely weak signal at 1,550 cm^{-1} (marked by an arrow) is the only visible signal due to the treatment. However, a difference spectrum (c) between spectra (a) and (b) indicates several absorption bands which coincide perfectly with the reference spectrum (e) of the preceding section. However, two broad bands seen in the 1,450 – 1,300 cm^{-1} and 1,200 – 800 cm^{-1} regions are not found in the reference spectrum. These bands are artifacts due to the imperfect subtraction of the quartz spectrum. Because different chemical treatment gives rise to different optical properties in the quartz, the spectrum of the quartz treated in a procedure using silane coupling reagent is slightly different from that treated without silane coupling reagent, leaving residual features due to the difference in quartz spectra. Similar to the preceding section, the cancellation of the quartz spectrum and its Restrahlen band is not easy. Thus the KBr pellet method described in this section and the transmission method explained in the previous section are not very useful for determining peak intensities in the range between 1,300 and 800 cm^{-1} in this sample.

1. N. Ikuta, T. Hirashima, E. Nishio, Y. Hamada, and Z. Maekawa, Preprints, 29th Dye Chemistry Annual Meeting (Tokyo), 68 (1987) (in Japanese).

Chart 4B

<Experimental conditions>
4 cm^{-1} resolution, DTGS detector, 36 scans coadded

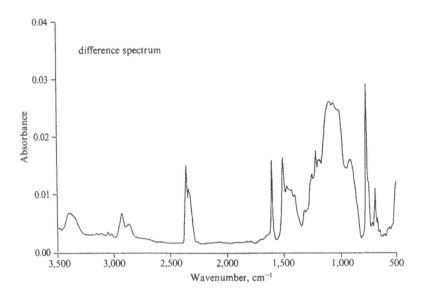

difference spectrum

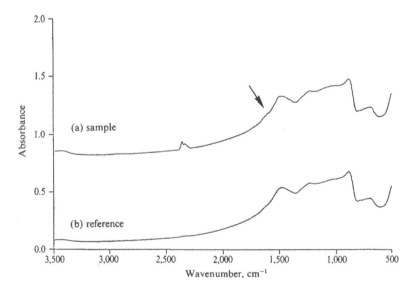

(a) sample

(b) reference

4C	Infrared Dichroism of Kevlar™ Fiber (I)	Microscope

Fibers are manufactured by extrusion from spinnerets.   Therefore, the polymer molecules of such fibers are more or less oriented.   When the degree of orientation is a measure of the production control and determines the physical properties of the fiber, it is important to determine the degree of polymer orientation in the fiber.   To do this without further processing the fiber, a dichroic determination of $10 \mu m$ diameter Kevlar™ single fiber using an IR microscope was attempted.

The aperture was set to cover the fiber $(10 \times 200 \mu m)$.   Since the draw axis lies on the fiber axis, parallel and perpendicular absorptions are measured when the electric field of IR radiation is parallel and perpendicular to the long axis of aperture, respectively.   Spectra (a) and (b) are the parallel and perpendicular absorption of a Kevlar™ single fiber, respectively.   The dichroic ratio of each absorption band was calculated using a baseline method in which an adequate straight line was drawn to connect the valleys on either side of peak.   Although distinct dichroism is obvious from the chart, dichroic ratios calculated from only the weak intensity peaks may be meaningful because the diffraction effect on the radiation by aperture and stray light[1] through the aperture is not negligible for strong absorption bands, as stated in Section 2.1.   Moreover correction of the dichroic ratio due to the wide range of different incident angles should be done using Eq. 2.1. Table 4.1 summarizes the dichroic ratio of the Kevlar™ single fiber studied here.   As shown in the table, the large dichroic ratio shows large correction, implying the number is not reliable. Thus, the orientation of molecules based on the dichroic spectra taken with a microscope must be considered as estimates.[2]

TABLE 4.1   Dichroic ratio of Kevlar™ fiber

Peak Position	$A_{//}$	$A_{\perp}$	R	R(corr)
1,150	0.54	0.14	3.9	4.1
1,116	0.53	0.27	2.0	2.1
1,026	0.71	0.07	10.0	11
984	0.23	0.06	3.8	4.0
900	0.49	0.31	1.6	1.7
863	0.28	0.43	0.65	0.64
823	0.37	0.56	0.66	0.65
790	0.50	0.03	15	19
734	0.42	0.20	2.1	2.2

[1] Correction of microscope dichroic ratio due to stray light has been discussed by Chase in: R. G. Messerschmidt and M. A. Harthcock ed., *Infrared Microspectropcopy*, pp. 93–102 (7. Dichroic infrared spectroscopy with microscope, B. Chase), Marcel Dekker (1988).

[2] It is possible to determine the degree of orientation in the case of PET fibers within 5% of standard deviation by the microscope method described in this section.

Chart 4C

<Experimental conditions>
4 cm^{-1} resolution, IR-PLAN microscope with medium band MCT detector, 100 scans

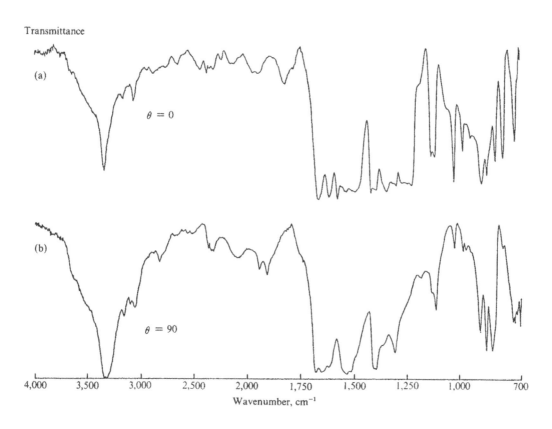

| 4D | Infrared Dichroism of Kevlar™ Fiber (II) | PAS |

Orientation of Kevlar™ molecules in the fiber was investigated by PA-FTIR method* and compared with the IR microscope study described in Section 4D.   It is possible to set a fiber across the sample cup (6 mm) by squeezing the fiber in the gap at both edges of the sample cup. Since the amount of sample is extremely small, $10\ \mu m\phi \times 6$ mm, compared with the reference carbon powder in the sample cup during acquisition of the background in the case of the standard PA-FTIR measurement procedure, the extra signal from the fiber will be diluted by the strong signal of carbon powder.   Therefore, single beam spectra of a $10\ \mu m$ diameter Kevlar™ fiber in the empty sample cup was measured using parallel and perpendicular polarized radiation.   A weak single beam PA-FTIR spectrum of the empty sample cup was subtracted from the sample single beam spectrum to cancel the signal due to sample cup.

Spectra in the chart are (a) parallel (E // fiber axis) and (b) perpendicular ($E \perp$ fiber axis) PA spectra of the Kevlar™ single fiber, respectively.   Since single beam PA-FTIR intensity has a unit of *absorbance* x *IR radiation energy* as discussed in Section 2.3, the dichroic ratio was calculated using the peak intensities of the dichroic spectra.   This process will automatically remove the influence of the IR source intensity.   The dichroic ratio of several absorption bands are listed in Table 4.2.   Similar to the preceding section, a baseline to connect the minima at high and low frequency sides of a peak was drawn to calculate the peak intensity.

Dichroic ratios obtained in this study show the same relationship with those observed with the IR microscope discussed in the preceding section.   Although stray light correction and correction for wide range incidence is not necessary in the case of PA-FTIR spectra, PA spectra are easily saturated.   Thus, one must be cautions in utilizing the PA-FTIR technique for quantitative determination of dichroic ratio.

TABLE 4.2

Peak Position (cm⁻¹)	Photoacoustic Signal I		Dichroic Ratio $R(I_{//}/I_{\perp})$
	parallel	perpendicular	
3,337	65	95	0.68
1,303	5.5	30	0.18
1,114	77	55	1.4
1,020	80	20	4.0
984	33	17	1.9
897	98	61	1.6
865	71	73	0.97
825	65	92	0.71
790	60	9	6.7
730	69	42	1.6
680	25	41	0.61
529	43	71	0.61
446	56	0	∞

* Data supplied by Dr. J. McClelland of MTEC.

Chart 4D

<Experimental conditions>
8 cm^{-1} resolution, MTEC PAS detector model 200, KRS-5/Al wire grid polarizer, OPD velocity = 0.05 cm s^{-1}, 160 scans

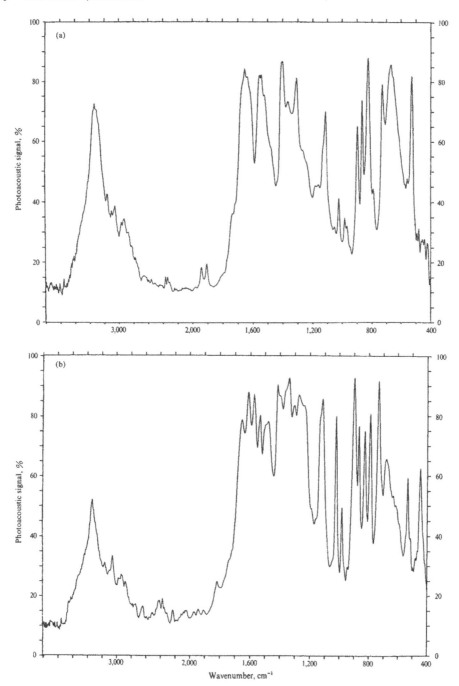

4E	**Surface Treatment of Fiber**	**ATR**

A sampling technique for the ATR study of fibers is presented in this section.   Although one may cut the fiber into short pieces and press the pieces against the IRE, the best method is to wind the fiber as densely as possible, as shown in Fig. 4.2, on a piece of hard board.   In this procedure one must avoid crossing of fiber.   This method is preferred because the experimental conditions are reproducible and the best contact of fiber with the IRE is achieved.

Elastic polyurethane fiber used in the clothing industry was analyzed.   Since the human body perspires, the surface of polyurethane fiber is treated with chemicals to prevent degradation of the fiber.   Using the above-mentioned sampling technique, we determined the surface treatment materials.

Figures (a) and (b) in the chart are ATR spectra of surface treated and untreated polyurethane fibers.   Difference spectrum (c) in absorbance units is also presented.   Since the treatment is so sparse, difference in spectra (a) and (b) is virtually invisible.   However, the difference spectrum (c) shows four characteristic absorption bands attributable to poly(methyl siloxane), or silicone, as marked by an asterisk (*) at 1,265, 1,100, 1,020, and 815 cm^{-1}.   Two broad bands marked with a dagger (†) at 1,470 and 1,380 cm^{-1} are due to CH_2 and CH_3 bending vibrations and are typical of mineral oil.   Thus it was concluded that the fiber is treated with a mineral oil solution of silicone oil.   Imperfections in the difference spectrum (c) are due largely to the bilayer structure of the fiber as explained in Section 2.2.

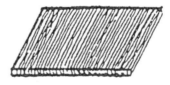

Fig. 4.2 ATR sampling method for fibers.

Chart 4E

<Experimental conditions>
4 cm^{-1} resolution, DTGS detector, 100 scans ATR conditions: KRS-5 IRE, 45° angle of incidence

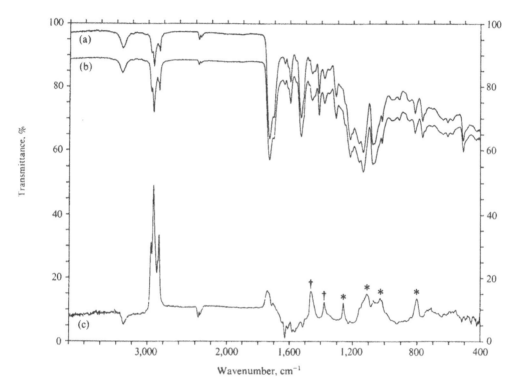

Ordinate values of spectrum (b) offset.   Spectrum (c) in absorbance

4F	Hair Treatment Reagent	Micro-ATR

A hair treatment reagent left on the surface of human hair was examined by ATR technique. Since it is desirable not to request a large amount of sample, a micro-ATR technique was tried, using a single strand of hair.   The chart shows a micro-ATR spectrum of a female hair treated with conditioners, including one formulated to prevent split ends (a) and a strand of hair from a male without any treatment (b).   At the bottom, the difference spectrum (c) between (a) and (b) is shown in absorbance units.

As seen in spectrum (a), a sharp peak at $1,260$ cm^{-1} and a broad doublet feature at $1,130$ and $1,050$ cm^{-1} suggest the presence of poly(methyl siloxane) or silicone oil.   In the difference spectrum (c), the major feature is the IR spectrum of the silicone oil.   Thus, the conditioner contained silicone oil and this silicone oil was left adsorbed on the surface of the hair.

This experiment demonstrated that a micro ATR technique removes the necessity of the laborious mounting of fibers in the ATR accessory.

Chart 4F

<Experimental conditions>
IR-PLAN with ATR objective 8 cm^{-1} resolution, narrow bond MCT detector, 64 scans

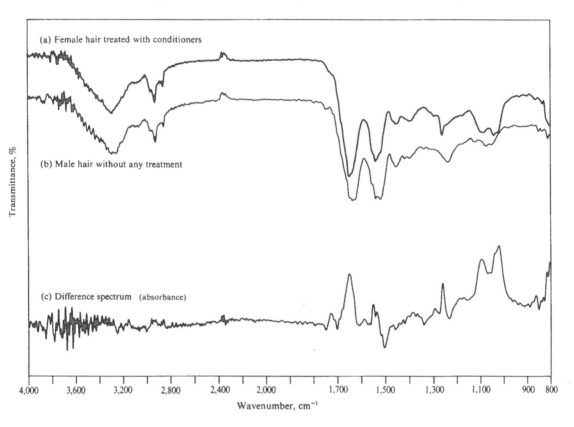

(a) Female hair treated with conditioners

(b) Male hair without any treatment

(c) Difference spectrum (absorbance)

Transmittance, %

Wavenumber, cm^{-1}

5. Powder and Bulky Samples

5.1 Diffuse Reflection

Diffusely reflected radiation contains a specular reflection component as discussed in Section 2.3.   The influence of the specular reflection component on diffuse reflection spectra will be examined, using fine powder of calcium carbonate, $CaCO_3$, as an example.   Spectrum (a) in the chart is a diffuse reflection spectrum of (neat) powdered $CaCO_3$.   Since the device to block the surface specular reflection component is part of the diffuse reflection accessory, the specular component from the very first layer of the powder particles is removed.   Spectrum (b) is measured from the diluted (0.5 wt%) $CaCO_3$ powder in KBr.   Positions, relative intensities, and linewidths of bands in spectrum (b) coincide qualitatively with those of the absorption spectrum cited in a standard reference library.   Although spectrum (a) agrees with the absorption spectrum for weak bands, the broad peak at *ca.* 1,400 cm^{-1} splits into two bands and the sharp band at *ca.* 800 cm^{-1} shows additional splitting.   Since the spectra are those of $CaCO_3$ in both cases, the cause of the splittings is attributed to an optical artifact which disappears on dilution.

Spectrum (c) is a difference spectrum between (a) and (b).   It coincides with the specular reflection spectrum of $CaCO_3$ crystal   Therefore, the splitting of the 1,400 cm^{-1} peak as well as that of the 880 cm^{-1} peak are attributed to the overlapping of the specular reflection component of the sample.   On the other hand, specular reflection components at 3,000 and 2,500 cm^{-1} are both weak where the anomalous dispersion of the refractive index around those frequencies is not conspicuous.   Thus, it must be remembered that strong absorption bands of some strongly absorbing inorganic compounds (TiO_2, Fe_2O_3, SiO_2, $BaSO_4$ and so on) show distortion in the diffuse reflection spectrum due to the overlapping specular reflection component.

In the case of organic compounds, however, since most of the absorption bands are weaker and narrower than those of inorganic compounds, an extraordinary effect such as shown in spectrum (a) in the 1,400 cm^{-1} region is not common.   Thus, in most cases, a diffuse reflection spectrum of a neat organic compound seems to coincide quite well with the absorption spectrum except for

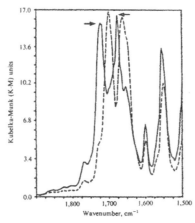

Fig. 5.1   Diffuse reflection spectra of neat powder of caffeine (solid line) and caffeine diluted (0.5%) in KBr powder (dotted line)

relative peak intensities.   However, one should not overlook the shifts of strong absorption lines. Fig. 5.1 exemplifies the shifts of two strong bands (marked by arrows) in the diffuse reflection spectrum of caffeine (solid line) compared with those of caffeine diluted (dotted line) in KBr powder obtained by diffuse reflection.   In addition, noticeable distortions will be observed even in the case of organic compounds when the particle size is large.   For instance, Fig. 5.2(a) shows a diffuse reflection spectrum of aspirin powder diluted in KBr powder (1 weight%).   Although

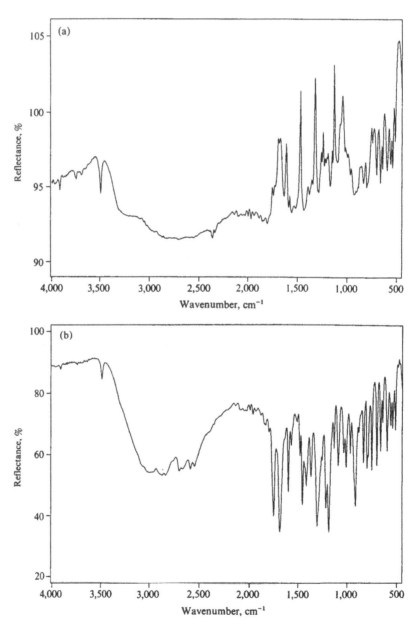

Fig. 5.2   Diffuse reflection spectra of 1 weight % granular aspirin in KBr powder (a) and 1 weight % ground aspirin in KBr powder (b).   (4 cm⁻¹ resolution, 36 scans, DTGS detector)

Chart 5.1

<Experimental conditions>
4 cm^{-1} resolution, DTGS detector, 16 scans

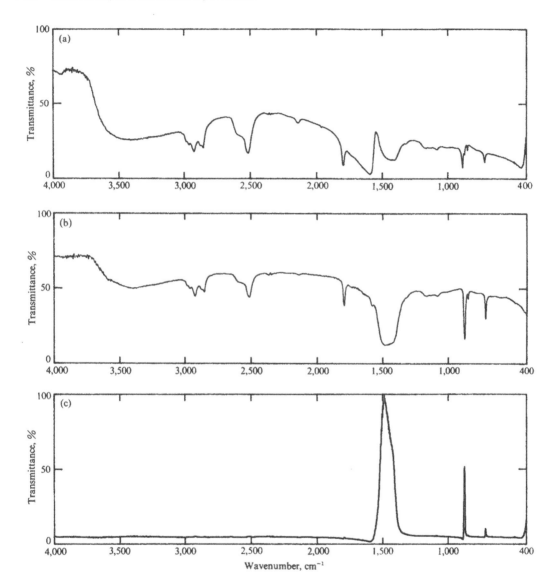

the aspirin sample looked like fine powder, the spectrum showed severe distortion.   We have found by optical microscope that aspirin particles are needle-like crystals with well-developed crystalline surfaces, and the size of these particles are *ca.* 15 micrometers in diameter ×30 – 80 micrometers long.   However, when the aspirin sample was ground and diluted in KBr powder (1 weight%), the diffuse reflection spectrum (Fig. 5.2(b)) showed normal absorption bands of sufficient intensity without such distortion.

5.1A | Diffuse Reflection of Silica (I). Particle Size

Substantial numbers of samples composed of or containing silica and silica gel are submitted to the IR analyst.   For instance, adsorbents and their reaction on silica-supported catalysts, organic compounds developed on the silica gel in thin layer chromatography (TLC) separation, Langmuir-Blodgett films formed on a fused silica plate, anti-corrosive protection layer of steel plate, silica film formed on a silicon-wafer, and surface-treated silica fiber and silica cloth are samples frequently studied by FT-IR spectroscopists, most of which are discussed in this volume. Let us explain how to obtain distortion-free diffuse reflection spectra of powders, utilizing a silica gel powder as an example.

As discussed in Section 2.3, (1) size of particle, (2) concentration of the sample, and (3) optical factor influence the quality of diffuse reflection spectra.   In this section, we will show the effect of particle size on the diffuse reflection spectrum of silica gel powder.

Spectra (a), (b), and (c) are the diffuse reflection spectra of 60, 120, and 230 mesh silica gel powders, respectively.   All of the samples have been equally diluted to 0.5 weight% in KBr powder.   In order to visualize the effect, 100 scans were coadded for all samples.   It is clear that the signal-to-noise ratio is improved as the size of the particles becomes small.   It is also clear that the diffuse reflectivity becomes higher as the particle size becomes smaller.   Moreover, spectral features such as the relative intensities of the major absorption bands and the shape of the Si-O stretching vibration showing the presence of a shoulder at the higher frequency side of the peak show more resemblance to the absorption spectrum of the silica gel as shown in Fig. 5.3, as the particle size becomes smaller.   Thus, the sample must be ground into a fine powder to obtain diffuse reflection spectra with minimal distortions comparable to absorption spectra.

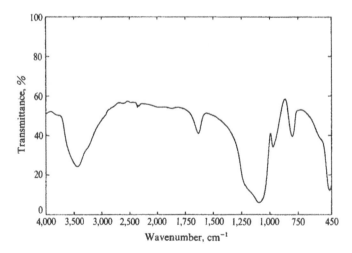

Fig. 5.3   Absorption spectrum of silica gel (KBr pellet).

Chart 5.1A

<Experimental conditions>
4 cm^{-1} resolution, DTGS detector, 100 scans, 16 scans for KBr pellet of silica gel.

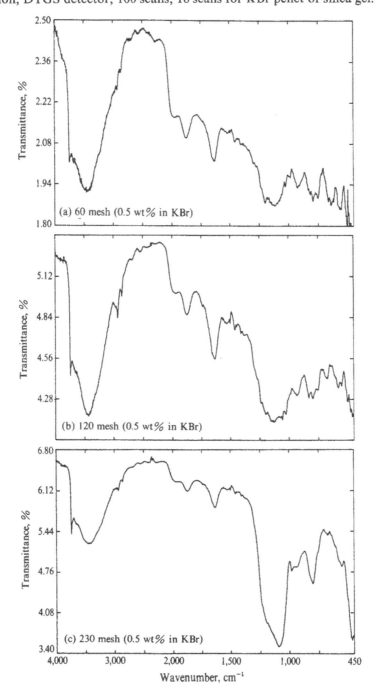

(a) 60 mesh (0.5 wt% in KBr)

(b) 120 mesh (0.5 wt% in KBr)

(c) 230 mesh (0.5 wt% in KBr)

Wavenumber, cm^{-1}

5.1B	Diffuse Reflection of Silica (II). Dilution Effect

In the preceding section, it was shown that the sample must be ground, using a mortar and pestle, as fine as possible to obtain spectra with less distortion and higher signal-to-noise. Generally speaking, the sample must be ground to such an extent that the particles start coagulating with each other (this phenomenon is called "secondary coagulation"). When the particles is ground to this level or smaller, the diffuse reflection spectra do not show any additional changes. As explained in Section 2.3, in order to remove the influence of the specular reflection component the sample powder must be diluted with a non-absorbing powder.

In this section, we will examine the effect of dilution on the shape of diffuse reflection spectra. Spectra in the chart are diffuse reflection spectra of (a) neat, (b) 10 weight% (in KBr), and (c) 1 weight% (in KBr) 230-mesh silica powders, respectively. Although distortion of the OH group absorption in the $3,800 - 3,000$ cm^{-1} region is not seen, there is one extra peak at around $1,350$ cm^{-1} in the case of the neat powder. This extra peak disappears when the powder is diluted in KBr powder at a concentration less than 10 weight%. The extra peak in the diffuse reflection spectrum silica powder arises for the same reason that of neat calcium carbonate arises, as discussed in Section 5.1; this extra peak is attributed to the artifact related to the *specular reflection* component of the Si–O vibration whose *absorption* component appears at about $1,100$ cm^{-1}. It should be noted that the artifact was not reduced at all by dilution to 50 weight%. Dilution down to the 1 weight% level gives rise to spectrum (c), whose spectral characteristics are close to an absorption spectrum, Fig. 5.2. However, unnecessary dilution will cause loss of signal from the sample and requires a large number of scans to recover the signal-to-noise.

Dilution effect on diffuse reflection spectra of organic compounds has been mentioned in Section 5.1, using diluted and neat caffeine as samples.

Chart 5.1B

<Experimental conditions>
4 cm^{-1} resolution, DTGS detector, 25 scans

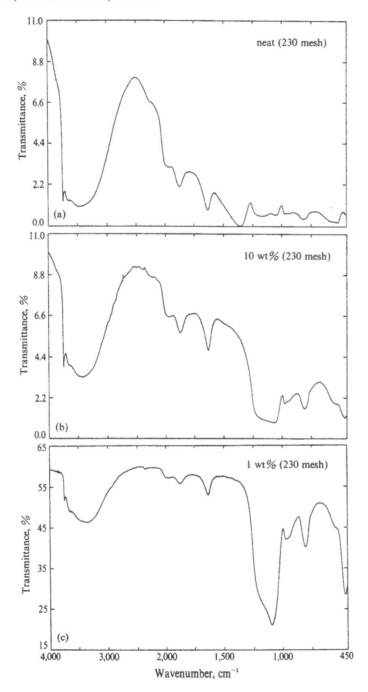

5.1C	IR Spectra of Adsorbent of Catalyst [1]	Diffuse Reflection

Silica and silica gel are frequently used as catalysts.   Already in the 1950s, IR spectra of a chemical species adsorbed on the surface of a catalyst were discussed in terms of the reaction mechanism, chemical structure of the adsorbents and the nature of the active sites of the catalyst.[2]   Although those pioneers exploited a transmission cell designed to allow the IR radiation to pass through powdered catalyst most of the time, the optical scheme of the transmission method is actually a pseudo-diffuse reflection measurement.   Most of the works conducted over the past ten years have been done with a diffuse reflection accessory equipped with a vacuum cell with heating capability.[3]   Two such examples are presented in this section.

As a model of amine adsorption to acidic sites of a silica catalyst, silica gel powder treated with aqueous ammonia was examined.   After NH_3 molecules were chemically adsorbed at the acidic sites of the silica, the silica was evacuated in order to remove water and non-adsorbed ammonia. A series of spectra, 1 through 6 in chart (a), show the Kubelka-Munk (KM) diffuse reflection spectra of the sample measured at room temperature after the sample was further heated under vacuum for 10 minutes at (1) 150°C, (2) 200°C, (3) 250°C, (4) 300°C, (5) 350°C, and (6) 400°C, respectively.   As shown by the arrow in the chart, peak intensities due to adsorbed NH_3 molecules decrease after treatment at elevated temperatures.

Chart (b) shows how the adsorbed carbon monoxide (CO) molecules disappear from the surface of the activated silica gel catalyst by the introduction of oxygen gas to the diffuse reflection vacuum chamber.   Repeated observations of the Kubelka-Munk corrected diffuse reflection spectra of the CO adsorbed on a silica gel catalyst were overlayed, for those cases where the catalyst was repeatedly (1 to 9) exposed to oxygen gas for 30 seconds followed by evacuation. The chart illustrates that the CO concentration decreases due to oxidation of adsorbed CO with oxygen to CO_2, which is not adsorbed on the catalyst.

---

[1] Data provided by Prof. Yoshiya Kera of Kinki University, Osaka, Japan.

[2] The technique to study catalysts by IR method was pioneered by Dr. R.P. Eischens of Texaco Beacon, U.S.A. and used in studies on absorbed species by Prof. Kozo Hirota, Professor Emeritus,

[3] Accessory available from Spectra-Tech Co., Harrick Scientific, and SPECAC Inc.

Chart 5.1C

<Experimental conditions>
4 cm^{-1} resolution, DTGS detector, 100 scans, Spectra-Tech Diffuse Reflection Accessory
(DRIFTS) with high vacuum cell, catalyst dispersed in KBr powder

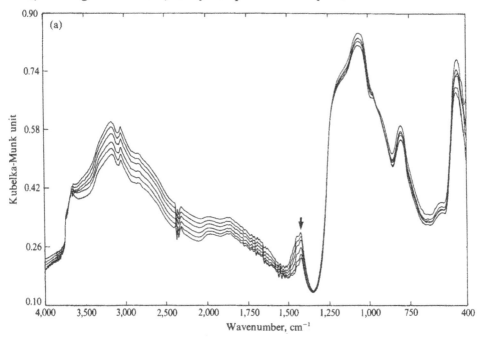

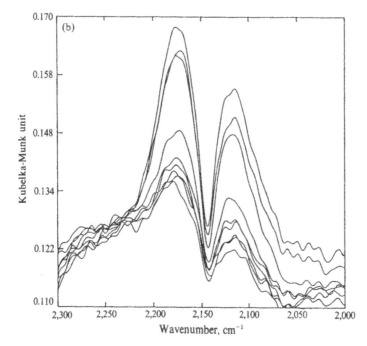

5.1D	Silanol Band of Silica Gel	Diffuse Reflection

A high-vacuum high-temperature cell was developed[1] for the diffuse reflection study of catalysis.   This type of cell allows for sample treatment with active and/or inactive gas at temperatures as high as $1,000°C$.   The data* shown here are an example of how the cell is applied in the case of silica gel.   As shown in the Kubleka-Munk (K-M) spectrum (a) of the chart, silica gel shows a broad absorption due to water molecules in the gel at around $3,350$ cm^{-1}.   However, a weak peak at $3,750$ cm^{-1} assigned to the Si–OH group is also observed.   Spectrum (b) was taken from the same silica gel heated at $1,000°C$ for 10 minutes under vacuum and cooled to room temperature while maintaining the vacuum.   In spectrum (b) the absorption band due to water is lost and a sharp absorption due to silanol is observed instead.

The silanol group is a chemically reactive functional group in silica gel, and chemical modification of the silica gel can be done through the reaction between the silanol group and a chemical reagent to modify the surface.   It was found that the level of the silanol group changes with temperature and period of thermal treatment.   In addition, the thermal treatment of silica at $1,000°C$ removes the water which interferes with the precise determination of silanol groups. If the precise concentration of the silanol group is known, it is possible to predict the degree of chemical modification or, in other words, the physical and/or chemical properties of the modified silica.

As shown in this experiment, the high-temperature high-vacuum cell for a diffuse reflection accessory becomes more versatile for studies of catalysis when the maximum temperature is higher.

1.   K. Nishikida, unpublished accessory and data.

* Experimental data courtesy of Mr. Senya Inoue and Mr. Junji Furuhashi of Kanto Kagaku Co., Ltd., Japan.

Chart 5.1D

<Experimental conditions>
8 cm^{-1} resolution, DTGS detector, 100 scans

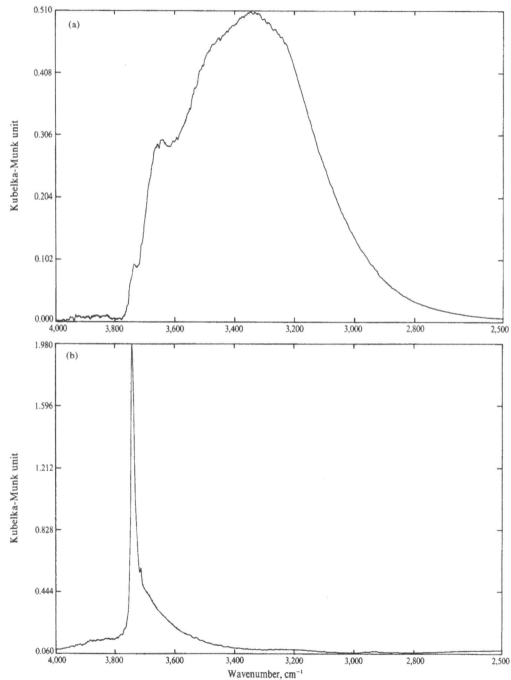

| 5.1E | Spot of Thin Layer Chromatograph | Diffuse Reflection |

Thin Layer Chromatography (TLC) is a technique frequently used by organic chemists, biochemists, and medical chemists to follow the synthesis and mutation of reagents and medicines.   A small portion of reaction mixture or solution of chemicals is placed on the TLC plate covered with silica or alumina.   After the solvent has evaporated, each component of the mixed chemicals placed on a position of the TLC plate is carried to different positions by developing the plate with various solvents or mixtures of solvents, thus separating each component into its own spot.   In order to identify the chemicals at these spots, each spot is scraped from the base plate and measured for diffuse reflection with or without dilution in KBr or KCl powder.

In this section, 2 μg of an experimental antibiotic compound was used to illustrate the features of the TLC method in FT-IR spectroscopy.   In the chart spectra (a) and (b) are the Kubelka-Munk spectra of silica gel taken from the spot and the area outside the spot, respectively.   The difference spectrum (c) between spectra (a) and (b) shows artifacts in the $1,300 - 1,100$ cm^{-1} and $550 - 450$ cm^{-1} regions in addition to several absorption bands of the substrate such as an intense line at $1,740$ cm^{-1} assigned to a carbonyl group and a band at $1,380$ cm^{-1} due to CH$_2$ and CH$_3$ groups.   The difference spectrum is not of sufficient quality to allow discussion of the structure of the compound.   Since the removed spot of silica powders was diluted to $ca.$ 1 wt% in KBr powder, the artifacts seen in spectrum (a) of Section 5.1 and spectrum (a) of Section 5.1B should not be observed in either spectrum (a) or (b) of this section.   As explained in Section 2.3, the optical process of diffuse reaction includes both refraction and reflection.   Thus, the diffuse reflection spectrum observed even from a properly diluted sample contains a small specular reflection component, leaving an artifact in the difference spectrum intense enough to jeopardize the study.   It should be remembered that the amplitude of the artifact in difference spectrum (c) is as large as the strongest absorption band of a 2-μg sample, which is far larger than the usual TLC experimental condition.   The best technique to observe the IR spectra of the compound from a TLC spot is a traditional method using the Wick-Stick method.*   Raman spectroscopy is also a satisfactory technique, since silica is a weak scatter and the Raman spectra of the TLC spot are free from silica interference.   However, the Raman method is not sensitive, and it requires a high concentration of analyte, $e.g.$ a few micrograms per spot.

---

*   A triangular stick ($ca.$ 6 mm(w)$\times$1 mm(d)$\times$15 mm(h)) made from KBr powder is held in the holder placed inside a small vial, at the bottom of which the substrate of the TLC plate was transferred from the spot.   When a small portion of a non-aqueous solvent such as chloroform is added, the solvent rises to the top of the KBr stick and evaporates. When all of the solvent is evaporated, the material on the spot is concentrated at the top of the triangular stick, the Wick-Stick.   A 1 mmϕ KBr pellet is made from the top portion of the stick to measure the absorption spectrum.

Chart 5.1E

<Experimental conditions>
8 cm^{-1} resolution, DTGS detector, 30 scans

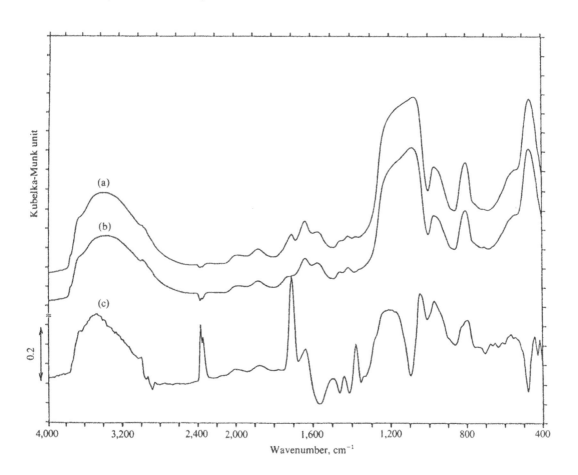

5.1F	Analysis of Liquid Chromatography Effluent	Diffuse Reflection

The diffuse reflection method is a convenient tool to observe IR spectra of compounds separated by high performance liquid chromatography (HPLC).   The LC effluent is dropped on fine KBr or KCl powder in a sample cup of the diffuse reflection accessory at a time corresponding to the retention time of each component.   When the solvent has evaporated, the diffuse reflection spectrum can be measured.   However, this method cannot be applied to all LC effluent.   For instance, attempts on certain solvents containing water and buffer solutions would not be successful.

The chart displays an example of the separation of *cis*- and *trans*-isomers of terpene on HPLC. Three-microliter portions of HPLC effluent were dropped in sample cups filled with KBr powder. After the solvent, chloroform, evaporated, diffuse reflection measurements were perfomed for the two isomers.   Diffuse reflection spectra thus obtained showed a fairly good signal-to-noise, enough to render possible detailed discussion on the chemical structures of the sample.

Chart 5.1F

<Experimental conditions>
TGS detector, 100 scans

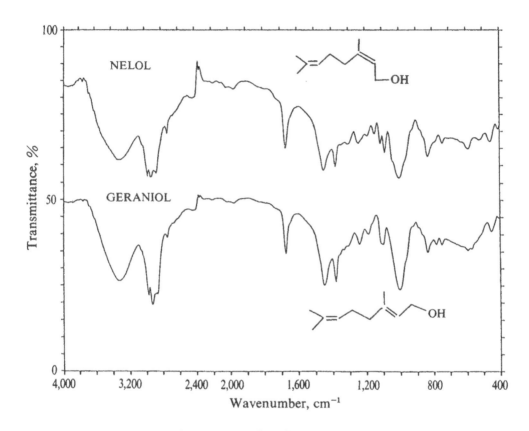

(Ordinate values offset for GERANIOL)

5.1G	Analysis of Photocopier Toner	Diffuse Reflection

Toners of photocopiers contain large amounts of carbon powder.   Since the carbon absorbs IR radiation extensively, only a small portion of toner should be added in KBr powder to avoid complete absorption on the IR radiation if the KBr pellet technique is being used.   However, since a sufficient concentration is needed to observe the absorption spectrum of the organic component, the KBr method requires trial-and-error to find the appropriate concentration of toner.   In contrast, it was found that the diffuse reflection method does not require any trial-and-error for most toner analyses.   The spectrum in the chart was observed from a neat toner without dilution, using the single beam spectrum of KBr powder as the background spectrum in the diffuse reflection method.

Peaks labeled with the letter S are due to some substituted benzene compounds and are assigned to polystyrene type resin, based on comparison with a standard spectrum cited in the library.   Peaks marked M are assigned to an ester compound.   Since styrene-methyl methacrylate copolymer is known to be used in photocopier toners, it is reasonable to assign the spectrum to styrene-methyl methacrylate copolymer.   A good match with a standard spectrum was found to confirm the compound as styrene-methyl methacrylate copolymer.

A strong peak was observed at 2,096 cm^{-1}.   Based on its peak position, the absorption band can be assigned to some salt consisting of CN–, SCN–, or NCS–.   However, it is not possible to make an unequivocal assignment.   We tentatively assigned it to Prussian blue pigment as discussed in the cases of Sections 3.2E and 3.3A.

The diffuse reflection spectrum of undiluted toner did not show any distortion because the toner is a fine powder and the organic compound is diluted by carbon powder.

Chart 5.1G

<Experimental conditions>
4 cm^{-1} resolution, DTGS detector, 50 scans

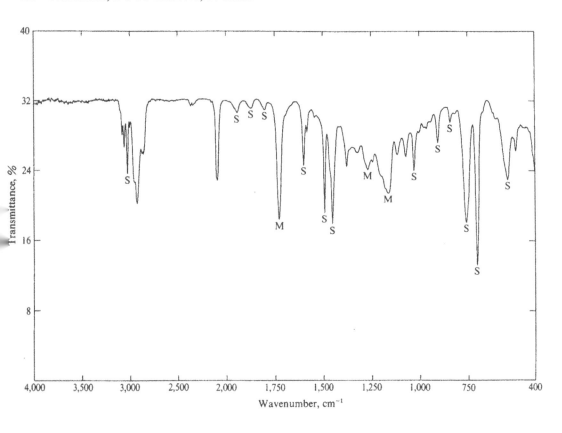

5.1H	Quantitative Analysis	Kubelka-Munk correction

In this section, we will show the necessity of Kubelka-Munk correction for quantitative analysis using diffuse reflection spectra.   In Section 2.4, it was explained that the optical pathlength is not uniquely determined in the case of diffuse reflection spectra.   This will result in erroneous quantitative determinations when diffuse reflection spectra are converted into absorbance.   In order to avoid this error, the diffuse reflection spectra must be converted into Kubelka-Munk units.

We diluted poly(methyl methacrylate) (PMMA) powder into barium sulphate powder.   The concentration of PMMA in the powder samples was 0 to 0.3% and spectra of neat sulphate and 5% PMMA in the sulphate are shown in Fig. 5.4.

The quantitation of PMMA in the sulphate was examined using a 1,740 cm^{-1} peak due to carbonyl stretching band of PMMA as an analyte line, since there is no interference from sulphate.   Kubelka-Munk-corrected spectra of different PMMA concentrations have been stack-plotted in the chart (a).   When the diffuse reflection spectra were converted into absorbance units, a curved relationship between peak height and concentration was obtained, as shown in the chart (b).   On the other hand, Kubelka-Munk spectra gave a linear relationship between peak height and concentration, the relationship being shown the chart (c).   It is clear that the Kubelka-Munk correction is required to obtain a linear calibration for quantitative analysis using the diffuse reflection technique.

On the other hand, when the concentration of the analyte is small or the peak intensity is weak, the absorbance unit may give a linear calibration.   In fact, almost all near-infrared (NIR) analyses using diffuse reflection technique are performed without Kubelka-Munk calculation.

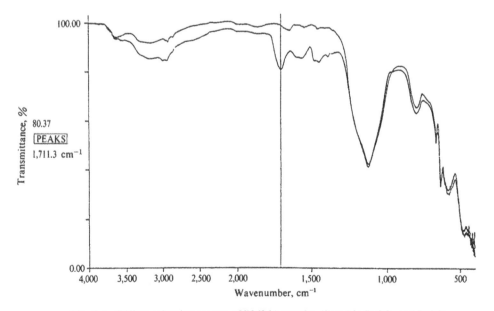

Fig. 5.4   Diffuse reflection spectra of PMMA powder diluted in BaSO₄ and BaSO₄.

Chart 5.1H

<Experimental conditions>
8 cm^{-1} resolution, DTGS detector, 36 scans coadded

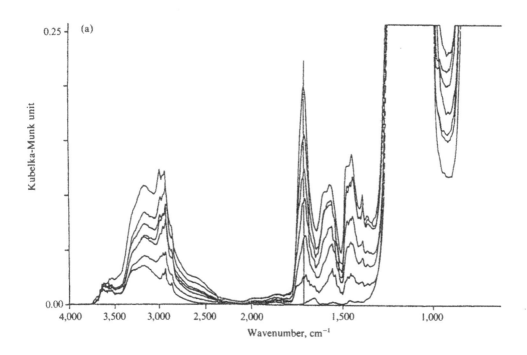

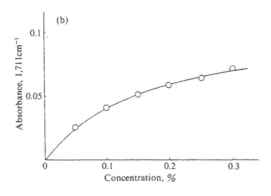

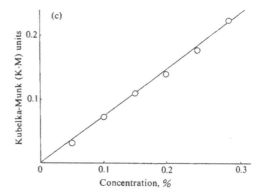

5.2   Photoacoustic-FTIR (PA-FTIR) Technique

5.2A	Surface Coating of Silica Gel [1)	PAS

Small particles of silica gel were coated with thin film of α-keratin,[*1] which is a native form of wool protein, the purpose of the study being to develop a model liquid chromatography column material to separate L- and D-isomers of natural products.   It became obvious that a good IR technique was necessary to obtain the spectra of the thin coating under the conditional of interference from intense absorption bands of silica gel.   This section concerns the merits of the PA-FTIR method for surface analysis.   Silica gel powder was first treated with a silane coupling reagent,[*2] so that the keratin was chemically bound to silica gel bridged by the silane coupling reagent.   Free and physically bound keratin were washed away.

Spectra (a) and (b) in the chart are PA-FTIR spectra of silica gel powder with and without coating, respectively, using carbon powder as the reference material.   Spectrum (a) shows a weak amide I band at $1,650\,cm^{-1}$, confirming the presence of the keratin coating layer. Difference spectrum (c) was calculated from spectra (a) and (b).   (Note that the PA-FTIR spectrum is already in units linearly related to concentration, and a data handling program to change the %T into absorbance should not be applied to the spectrum in the difference calculation.)   The amide I band in the difference spectrum agrees quite well with that obtained from the transmission spectrum of α-keratin, as shown in Fig. 5.5, suggesting that the keratin maintains the α-helix structure on the silica gel surface.

It should be noted that an attempt to obtain the IR spectrum of the coating layer by diffuse reflection technique was frustrated because of overwhelmingly strong silica gel absorption and an artifact due to the specular reflection component.   Although the optical process of the diffuse reflection phenomenon involves the transmission of the IR radiation through the entire diameter of the particle, the photoacoustic phenomenon does not require the contribution of the entire particle to the signal, as the modulation frequency determines the penetration.   Thus, PA-FTIR is more advantageous on surface analysis than the diffuse reflection method.

1)   E. Nishio, I. Abe, N. Ikuta, J. Koga, H. Okabayashi and K. Nishikida, *Appl. Spectr.*, **45**, 496 (1991).

*1 SCMKA: *S*-carbomethylated keratin
*2 McPS: γ-mercaptopropyltrimethoxy silane

Chart 5.2A

<Experimental conditions>
MTEC photoacoustic detector, 8 cm^{-1} resolution, 10 scans

PA-FTIR spectra of silica gel treated with SCMKA and McPS(a) and with McPS only (b).

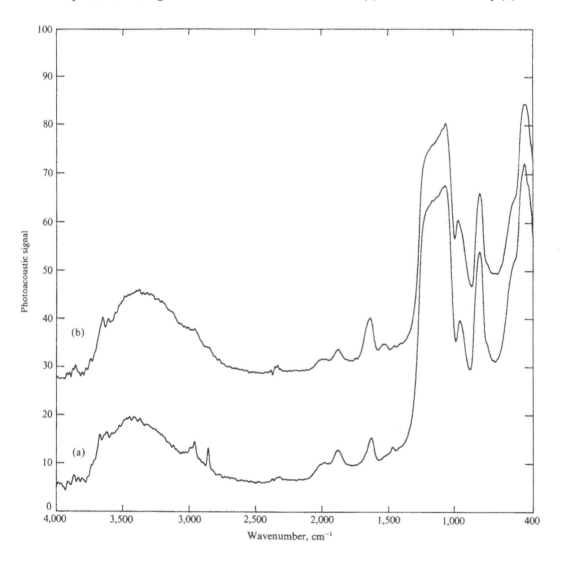

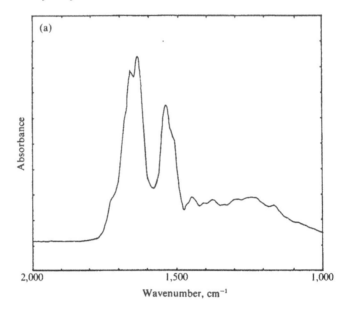

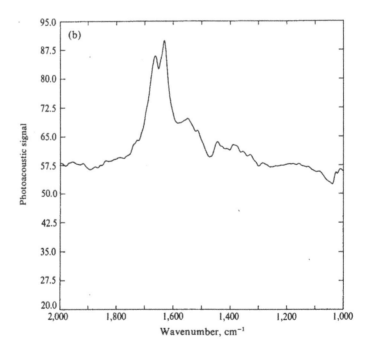

Fig. 5.5   Amide I and II bands of transmission spectrum of native keratin film and difference PA-FTIR spectrum between those of silica gel treated with SCMKA and McPS (a) and with McPS only (b).

5.2B	Surface Coating of Powder Detergent	PAS

Washing detergent for domestic use is a dry powder when the container box is newly opened. However, once the box is open, the detergent gradually coagulates and finally becomes a solid bulk.   Since coagulation occurs when the surface of the detergent particle starts having affinity to other particles, it is essential to investigate the difference between the surface of fresh and old detergent particles.

PA-FTIR spectra of fresh and old detergent were measured at four different OPD velocities, using carbon backgrounds with matched OPD velocities.   Charts (a) and (b) are PA-FTIR spectra of the fresh and old detergent measured with OPD velocities of 0.05 (1), 0.25 (2), 0.50 (3), and 0.75 (4) cm s^{-1}, respectively.   Remember that as the velocity decreases, the apparent sample thicknees increases.   The most prominent difference in the spectra of fresh and old detergent lies in the intensity of the peak at 1,000 cm^{-1}.   The thermal diffusion length for the 1,000 cm^{-1} radiation is calculated using the thermal diffusivity of sodium salt of the fatty acid equal to 1.3×10^{-3} cm^2 s^{-1}.   Resulting values are given in Table 5.1.   The intensity of the 1,000 cm^{-1} peak became relatively stronger when the shallow surface of the fresh detergent was measured. On the other hand, the intensity of the 1,000 cm^{-1} peak of the old detergent is weak compared with that of the fresh detergent and the relative intensity does not increase as dramatically as that of the fresh one as the shallow surface is measured.   These results indicate that a compound which shows a peak at 1,000 cm^{-1} is concentrated at the surface of the fresh detergent, while the distribution of this compound becomes more or less equal throughout the old detergent particles. Reviewing these experimental results, it may be concluded that the surface of the detergent particle is treated with a compound to keep the particle dry, preventing coagulation.   However, the compound gradually diffuses into the inner part of the particle, so that the concentration of the compound becomes nearly uniform throughout the particle, and at the same time the particle loses its anti-coagulation property.   The compound used for surface treatment shows intense absorption at 1,000 cm^{-1}.   Three compounds used in detergents are known to have strong absorption at around 1,000 cm^{-1}; these include alkyl-poly(oxylethylene) phosphate, silicate, and cellulose.   Although it is impossible to draw a definitive conclusion from the above spectra regarding which compound is used as the coating material, the manufacturer* of the detergent disclosed that the compound used for the surface protection was a specially manufactured zeolite, which is a kind of silicate.   As demonstrated by this experiment, PA-FTIR is a satisfactory method for the study of rather thick surfaces, as shown in Table 5.1.

TABLE 5.1   Relationship between thermal diffusion lengths
and OPD velocity at 1,000 cm^{-1}

OPD velocity	Thermal diffusion length
0.05 cm s^{-1}	28.6 μm
0.25	25.7
0.5	9.1
0.75	7.4

Chart 5.2B

<Experimental conditions>
4 cm^{-1} resolution, MTEC PAS detector, 50 (for 0.05 cm s^{-1} OPD velocity) to 500 (for 0.75 cm s^{-1} OPD velocity) scans.   Both samples dried over phosphorus pentoxide.

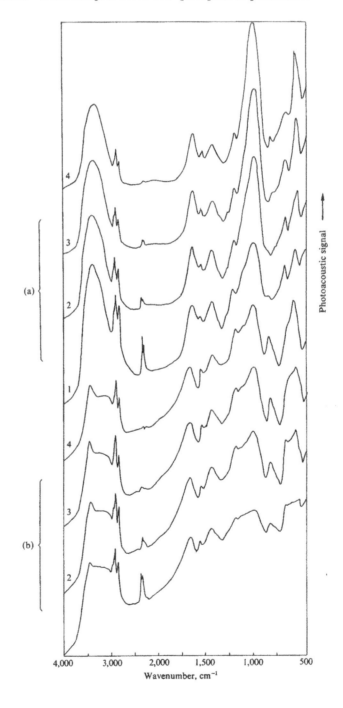

6.   Thin Films Formed on Solid Surfaces

6.1   Laminate Films on Metal Plates

Cans used for beer, soft drinks, and other drinks are made from aluminum or steel.   In order to avoid corrosion by the contents of the can, the inner side is coated with polymers.   These coatings are of single or multiple layers and the multiple layers are called "laminates."   In this section we will show analyses of polymer films of the inner surface of aluminum cans.

6.1A	Laminate Film on the Inner Surface of Alminum Can (I)	Specular Reflection

Since aluminum has high reflectivity, it is convenient to use the specular reflection accessory to obtain absorption spectrum.   This method is called "infrared reflection-absorption spectroscopy", or IR-RAS, as explained in Section 2.4.

Spectrum (a) of the chart is measured from the inner side of an aluminum can.   Using a diagram to identify polymer IR spectra shown in the appendix, one can easily identify it as an epichlorohydrin-type epoxide resin.   Spectrum (b) taken from the inner side of another kind of aluminum can shows a carbonyl band at $1,740$ cm^{-1} and all of the intense bands seen in spectrum (a), indicating that the materials used to make the coating are epoxy resin and another compound with a carbonyl group.   (Details of the laminate (b) will be discussed in the following section.)

It is important to observe the near infrared (NIR) region, $7,000 - 4,000$ cm^{-1}.   Although only weak absorption bands attributed to overtones and combination vibrations are observed in this frequency region, laminates usually exhibit a fringe pattern as marked by arrows.   Since the fringe pattern is caused by the interaction between the radiations reflected at the laminate surface and the aluminum plate, it is possible to calculate the thickness of the laminate, using Eq. 2.27. Using the frequencies at the arrows and a refractive index (assumed to be 1.5) of an organic compound, the thickness of laminates (a) and (b) was found to be 2.3 and 4.0 μm, respectively.

It is extremely interesting to realize that the can used for the same soft drink in the 1,970s was coated with a 12-μm thick epoxide resin.[1]   Because of advances in coating technology, the thickness has been reduced by three quarters, saving natural resources, at the same time providing the same anti-corrosion protection.

1.   R. W. Hannah, Infrared Bulletin No. 60, Perkin-Elmer Corporation.

Chart 6.1A

<Experimental conditions>
4 cm^{-1} resolution, DTGS detector, 16 scans, specular reflection accessory, 6.5° incident angle
with Al mirror as reference

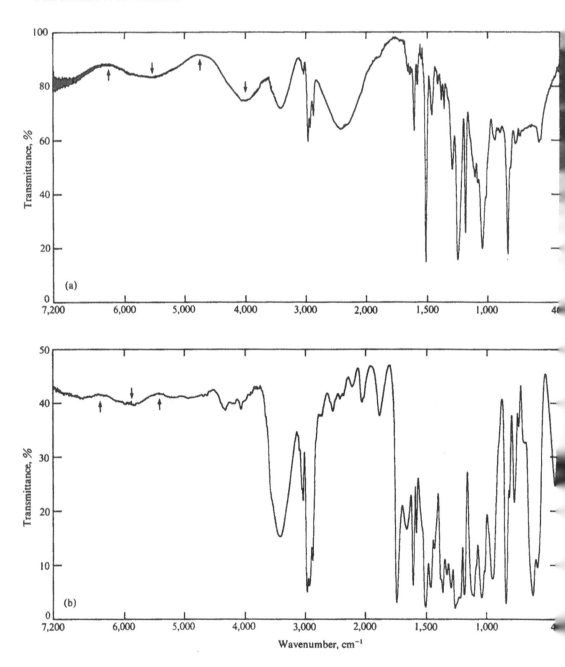

| 6.1B | Laminate Film on the Inner Surface of Alminum Can (II) | ATR |

One of the laminates examined in the preceding section, showed that it was composed of two compounds, epoxide and an ester-type compound. We decided to study this laminate with ATR, which allows observation of the surface to a depth of a fraction of the wavelength. Therefore, if the ATR spectrum shows the same composition as the IR-RAS showed, the laminate is made of the abovementioned polymer mixture. However, if the ATR spectrum shows the presence of only one of these compounds, the laminate will have a multilayer structure.

Spectrum (a) of the chart is the ATR spectrum of the laminate whose IR-RA spectrum was shown in spectrum (b) of the previous section. The spectrum of the epoxide resin seen in spectrum (b) of the previous section is almost completely lost from (a) of this section, and the spectrum shows the characteristic pattern of C–Cl vibrations of poly(vinyl chloride) near 600 cm^{-1}. Besides the presence of an ester group supported by a carbonyl band at $1,740 \text{ cm}^{-1}$, ester-type C–O vibration is seen at $1,240 \text{ cm}^{-1}$. Using the diagram for polymer assignment by IR spectra given in the appendix, it is reasonable that the compound be assigned to vinyl chloride-vinyl acetate copolymer. This is confirmed by comparison with a standard spectrum listed in commercial spectral libraries such as the Sadtler Library.

Thus, it was concluded that in the sample in question the epoxide resin was applid to an aluminum plate followed by the application of a vinyl chloride-vinyl acetate copolymer on the epoxide layer.

Chart 6.1B

<Experimental conditions>
4 cm^{-1} resolution, DTGS detector, 16 scans
ATR conditions: KRS-5 IRE, 45° angle of incidence

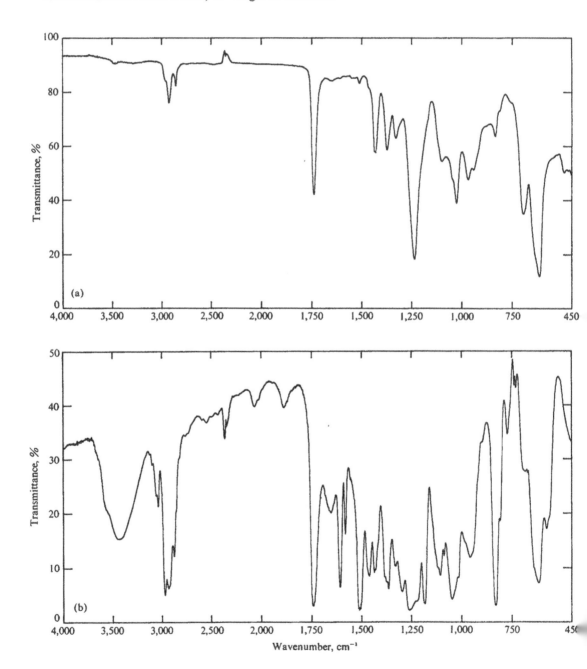

6.2 Langmuir-Blodgett (LB) Films

In this section, we discuss IR studies of LB films.[1] The importance of studies on LB film arises not only because of theoretical interest in its well-defined struture but also because of proposal regarding the possibility of developing molecular electronic devices from well-ordered organic compounds,[2] whereby the technique of LB film formation is the best suited. Although the possibility of achieving molecular electronic devices seems to be quite slim, biological and medical interest in LB film is growing.[3]

6.2A	LB Film on Metal Plate (I)	Grazing Angle IR-RAS

In this section, IR-RA spectrum of an LB film of a fatty acid salt accumulated on a silver plate is presented. The orientation of cadmium stearate molecules $(CH_3-(CH_2)_{16}-COO^-)_2Cd$ in LB film is well known. The stearate ion is anchored by silver metal at the COO^{-1} group, and the zigzag chain, $-(CH_2)_{16}-CH_3$, of the stearate lies along a line nearly normal to the silver metal surface. Each stearate ion is arranged to occupy a regularly repeated position, so that a monolayer of two-dimensional single crystals is formed as shown in Fig. 6.1(a). At temperatures higher than the transition point, the crystal structure of the CH_2 zigzag is lost, while the position of the COO^- group is not changed (Fig. 6.1(b)). It is also possible to accumulate many such layers on the first layer of a metal surface.

The chart represents the parallel polarized grazing angle (80° angle of incidence) IR-RA spectrum of a 9-layer (about 200 Å thick) LB film composed of cadmium stearate on silver metal substrate. Although the use of $(\Delta R/R)_{//}$, which is a theoretical ordinate value as described in Section 2.5, may be recommended, the use of the traditional absorbance unit calculated as $-\log(\Delta R/R)_{//}$ is justified, since $(\Delta R/R)_{//}$ is so small that the relative intensities of each line given as absorbance and $(\Delta R/R)_{//}$ values are almost the same.

Fig. 6.1 Schematic models for phases of LB film. (a) crystalline phase (b) liquid phase

1. a) F. Kimura, J. Umemura, T. Takenaka, *Langmuir*, **2**, 96 (1986); b) C. Naselli, J. F. Rabolt and J. D. Swallen, *J. Chem. Phys.*, **82**, 2136 (1985).
2. a) F.I. Carter, Problem and Prospects of Future Electroactive Polymess and "Molecular Electronic Devices" in the NRL Program on Electroactive Polymers, First Annual Repeat, (Ed., L.D. Lockhart, Jr.), *NRL Memorandom Report*, **3960**, P.121 (1979).
    b) Further Consideration on "Molecular" Electronic Devices. Second Annual Report, (Ed., R. B. Fox), *NRL Memorandom Report* **4335**, P. 35 (1980).
3. a) J. I. Anzai and T. Osa, *Selective Electrode Rev.*, **12**, 3 (1990); b) Y. Okahata, T. Tsuruta, K. Ijiro and K. Ariga, *Thin Solid Films*, **180**, 65 (1990); c) S. Arisawa and R. Yamamoto, *ibid.*, **210**, 443 (1992); d) D. H. Charych, J. O. Nagy, W. Spevak and M. D. Bednarski, *Science*, **261**, 585 (1993).

Chart 6.2A

<Experimental conditions>
4 cm^{-1} resolution, medium range MCT detector, 100 scans, AgBr wire-grid polarizer, variable angle specular reflection accessory

Conditions of LB film formation: ambient temperature, 15.6°C; pH value of water bath, 5.6; Cd^{2+} concentration in the water phase, 5×10^{-4} M; surface pressure, 30.8 dynes m^{-1}

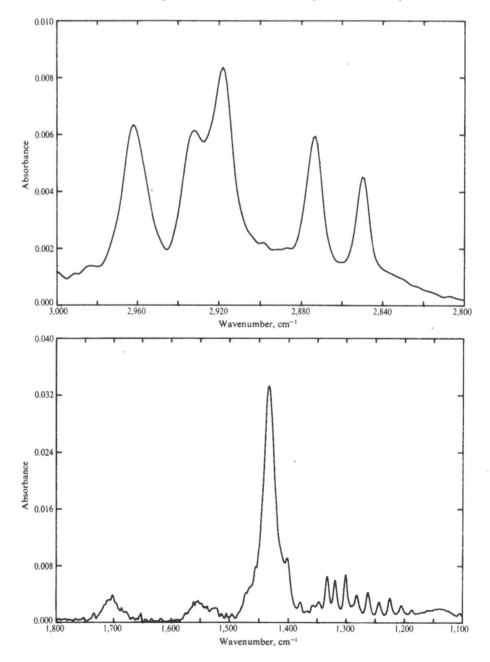

Five peaks in the CH stretching vibration region are due to methyl groups at 2,962, 2,932, and 2,873 cm^{-1} and methylene groups at 2,918 and 2,850 cm^{-1}. A weak band at 1,703 cm^{-1} is assigned to a carbonyl stretching vibration of stearic acid, showing that stearic acid coexists in the present LB film. A weak band at 1,561 cm^{-1} and a strong band at 1,433 cm^{-1} are assigned to anti-symmetric and symmetric stretching vibrations of the COO$^-$ group of stearic acid salt, $-COO^-M^+$, respectively. A characteristic band progreesion seen in 1,379 – 1,175 cm^{-1} region is assigned to highly coupled CH$_2$ wagging vibrations. The band progression appears in the IR spectra of the long chain compounds such as fatty acids, n-alkyl carbinols, and n-alkanes, when they are in a crystalline phase. All of the spectral features observed here agree well with the results reported in the literature.[1] The band progression is reported to decrease in intensity above the transition point because the line width of the individual fine structure becomes more disordered.

6.2B	LB Film on Metal Plate (II)	Grazing Angle IR-RAS

The chart presents a grazing angle IR–RA spectrum taken from a monolayer LB film on silver metal, the starting material and experimental conditions being the same as in the preceding section. Reflecting the fact that the film is thinner, the intensity of the spectrum is weaker than that of the nine-layer LB film. In addition, there is no positive sign of stearic acid, although the signal-to-noise is not good around the 1,700 cm^{-1} region for a definitive observation. The features of grazing angle IR-RAS spectra of thin film on metal plate are explained in this and in the previous section.

In the case of a transmission spectrum of unoriented stearic acid salt, the intensities of the two CH stretching vibration bands due to the CH$_2$ group are far stronger than the three CH stretching vibration bands of the CH$_3$ group, reflecting the ratio of 16 CH$_2$ groups to 1 CH$_3$ group. However. the intensities of the five bands are not much different from each other for monolayer and nine-layer LB films, as shown in the upper spectra of the charts in this and in the preceding section. In the case of the grazing angle IR-RAS, the electric field of the IR radiation available to interact with the film material at the surface is normal to the metal plate (z direction in Fig. 6.2(a)). When the directions (x-y plane of Fig. 6.2) of the transition dipole moments of the symmetric and anti-symmetric vibrations of the methylene group are nearly orthogonal to the electric field, the intensity of these vibration bands will be weak, $i.e.$ these vibrations are a forbidden transition. On the other hand, the C–H stretching vibration of the methyl group will be an allowed transition, because the directions of the transition moments of a freely rotating methyl group cannot be orthogonal to the abovementioned electric field. Thus, the relative intensities of the CH stretching vibration region are consistent with the orientation of the stearate ion described in the preceding section.

There is a similar explanation for the intensities of the two stretching vibrations of the COO$^-$ group. If a line connecting two oxygen atoms of the COO$^-$ group is parallel to the metal surface, the induced dipole moment is normal and parallel to the electric field of IR radiation for anti-symmetric and symmetric stretching vibrations respectively of the COO$^-$ group, as shown in

1. J. D. Swalen et $al.$, $Langmuir$, **3**, 932 (1987).

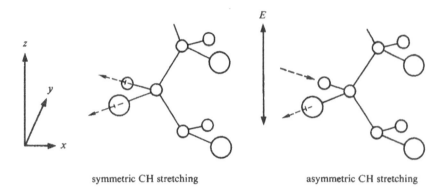

symmetric CH stretching   asymmetric CH stretching

Fig. 6.2   Relative orientations of surface electric field and transition moments of symmetric and anti-symmetric CH_2 stretching vibrations.

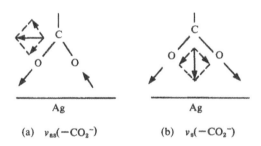

(a) $\nu_{as}(-CO_2^-)$ (b) $\nu_s(-CO_2^-)$

Fig. 6.3   Relative orientation of surface electric field and transition moments of anti-symmetric (a) and symmetric (b) COO^- group.

Fig. 6.3. Thus, the intensities of anti-symmetric and symmetric vibration become 0 and maximum, respectively.   Since the former vibration is observed to be of weak intensity while the intensity of the latter vibration is strong, the line connecting the two oxygens makes a small angle with the metal surface.   Allara[1] states that the angle is *ca*. 25° in the case of the hexadecanoic acid LB film.

Some of the lines in the CH_2 band progression are seen even in the spectrum of the monolayer, although a bent baseline prohibits an unambiguous assignment.

1.   D.L. Allara, R.G. Nuzzo, *Langmuir*, **1**, 52 (1985).

Chart 6.2B

<Experimental conditions>

4 cm^{-1} resolution, medium range MCT detector, 1,000 scans, AgBr wire-grid polarizer, variable angle reflection accessory with silver reference mirror

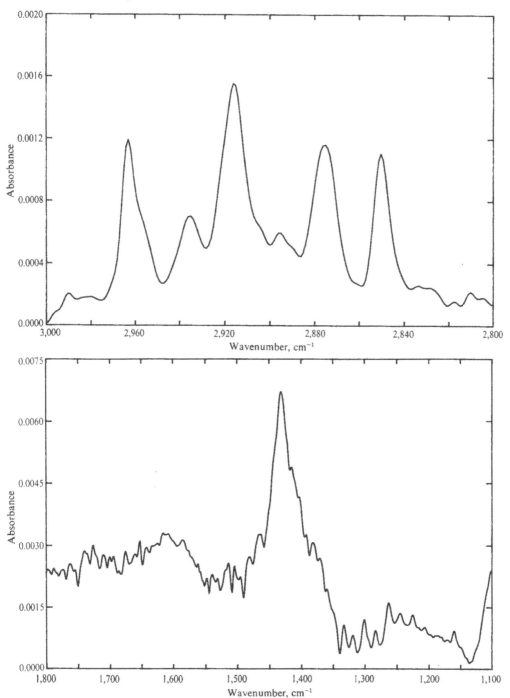

6.2C	LB Film on CaF₂ Plate	Transmission

Spectra (a) and (b) in the chart are transmission spectra taken from CaF_2 plates coated with a monolayer and nine layers of LB film, respectively, formed from cadmium arachidate (CH_3–$(CH_2)_{18}$–$COO^-)_2$ Cd.   Since the conditions for the formation of LB film are the same as those employed for LB film on silver metal, the orientation of the arachidic acid salt is assumed to be similar to that of stearic acid salt on metal substrate.   However, transmission spectra and grazing IR-RA spectra are quite different in the relative intensities of many bands.[1]   In this section, reasons for such differences are discussed.

When IR radiation travels in the z direction (normal to the CaF_2 plate), the electric field of unpolarized IR radiation lies in the x-y plane (in the CaF_2 plate), as shown in Fig. 6.4.   Since directions of the induced dipole moment for $\nu_{as}(CH_2)$ and $\nu_s(CH_2)$ vibrations lie in the x and y axes, respectively, the induced dipole moment can interact with the electric field, giving rise to allowed absorptions.   As discussed in the preceding section, the methyl group is freely rotating and again induced dipole moments can interact with the electric field.   Thus, roughly speaking, the relative intensities of the two CH_2 absorption bands and the three of the CH_3 group become close to those of an unoriented sample, reflecting the numbers of methylene and methyl groups in an arachidate ion.

The relative intensities of $\nu_{as}(COO^-)$ at $1,550\ cm^{-1}$ and $\nu_s(COO^-)$ at $1,430\ cm^{-1}$ are also reversed from those of the grazing angle IR–RA spectrum, that is, the $\nu_{as}(COO^-)$ is stronger than $\nu_s(COO^-)$ in the transmission spectra.   The reason for this reversal is also explained in terms of the directions of electric field of IR radiation and vibration-induced dipole moment.   The induced dipole moment of anti-symmetric vibration of the COO^- group is parallel to the line connecting two oxygen atoms and that of symmetric vibration is parallel to the line bisecting the angle, $\angle OCO$.   Therefore, if the line connecting the two oxygen atoms of the COO^- group is parallel to the CaF_2 plate, $\nu_{as}(COO^-)$ has the maximum intensity and $\nu_s(COO^-)$ has zero intensity.   Since the intensity of $\nu_s(COO^-)$ is not zero, the observed spectra support the hypothesis that the arachidic acid salt is slightly tilted from the normal to the CaF_2 plate, as discussed in the preceding section.

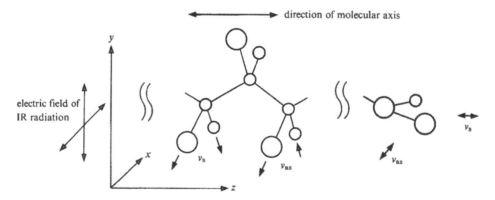

Fig. 6.4   Relationship between electric field of IR radiation and transition moments of CH_2 group (left) and COO^- group (right).

1.   C. Naselli, J.F. Roberts, and J.D. Swalen, *J. Chem. Phys.*, **82**, 2136 (1985).

A band at 1,460 cm^{-1} is assigned to the bending vibration of the CH$_2$ group.   A band progression in the region 1,400 – 1,200 cm^{-1} is also seen in the case of the nine-layer LB film.

Chart 6.2C

<Experimental conditions>
4 cm^{-1} resolution, medium band MCT detector, 100 scans for the nine-layer sample and 1,000 scans for the monolayer sample

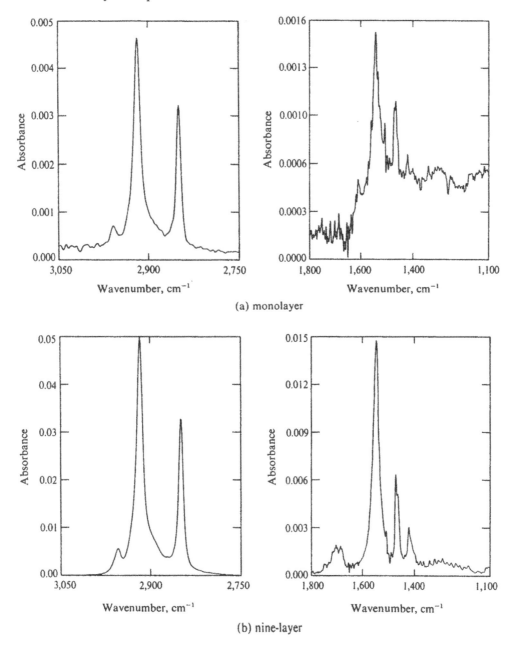

(a) monolayer

(b) nine-layer

6.2D	LB Film Formed on Silica Plate (I)	External Reflection

Reflection spectra of a silica plate coated with nine layers of LB film were measured with polarized light with incident angles varying from 25° to 75°.   The experimental arrangement for the present "Infrared External Reflection Spectroscopy (IR-ERS)" is exactly the same as that of the IR-RAS experiments discussed in Sections 6.2A and 6.2B.   Cadmium stearate was used to prepare the LB film.

The reflectivity of silica becomes considerable and similar to a metal plate at high incident angles.   Therefore, one may expect similar signal enhancement for the silica substrate as in the case of the grazing angle IR-RAS of metal-supported LB films.   However, as explaind in Section 2.3, IR-ERS is theoretically different[1),2)] from the grazing angle IR-RAS.   In this section, the difference between grazing angle IR-RAS and IR-ERS is discussed.

IR-ER spectra of above-mentioned silica measured with (a) 75° angle of incidence of s-polarized, (b) 45° angle incidence of p-polarized, and (c) 45° angle of incidence of p-polarized light are given in the chart.   While the main feature in each spectrum is the reflection spectrum of silica observed at each incident angle and polarization, weak spectral features due to the LB film are observed in the 3,000 – 2,800 cm^{-1} region as expended in the chart.   Moreover, a peak was observed at 1,538 cm^{-1}, corresponding to $\nu_{as}(COO^-)$ of arachidate salt.   Although the grazing angle IR-RAS with s-polarized light does not give any spectral features due to the thin film (see Section 6.3E), IR-ERS provides features due to the thin film component on the substrate as demonstrated in spectrum (a) and its expanded part.   Angular dependence of the peak intensity of the thin film component is also different between IR-ERS and grazing angle IR-RAS. Although the intensity of the thin film component reaches maximum at a grazing angle such as 80°–85° (see Section 6.3E), the optimum angle was around 55°[1)] for IR-ERS (see following section).

The reflectivity of silica in the non-absorbing frequency region becomes nearly zero at a 50° angle of incidence with p-polarized light, since this incident angle approaches the Brewster angle of silica, at which the reflectivity of substance becomes zero for p-polarized light.   Although it is difficult to obtain a good signal-to-noise, two derivative-like bands due to $\nu_{as}(CH_2)$ and $\nu_s(CH_2)$ were observed near 2,900 cm^{-1}, as shown in spectrum (c).   Tanaka[1)] and Allara[2)] observed derivative-like line shapes in certain cases of thin coating materials on nonmetallic substrates.   Also, Tanaka showed that the phase of the $\nu_{as}(CH_2)$ and $\nu_s(CH_2)$ changes at the Brewster angle.   So far as we know, there is no derivative-like spectra in the mid-IR region observed from the thin film on *metal* substrate with grazing angle IR-RAS.

1.   A. Udagawa, T. Matsui, and S. Tanaka, *Appl. Spectrosc.*, **40**, 794 (1986).
2.   M.D. Porter, T.B. Bright, D.A. Allara, and T. Kuwana, *Anal. Chem.*, **58**, 2461 (1986).

Chart 6.2D

<Experiment conditions>
8 cm⁻¹ resolution, DTGS detector, 100 scans
SPECAC "Monolayer" variable angle specular reflection accessory with AgBr wire-grid polarizer

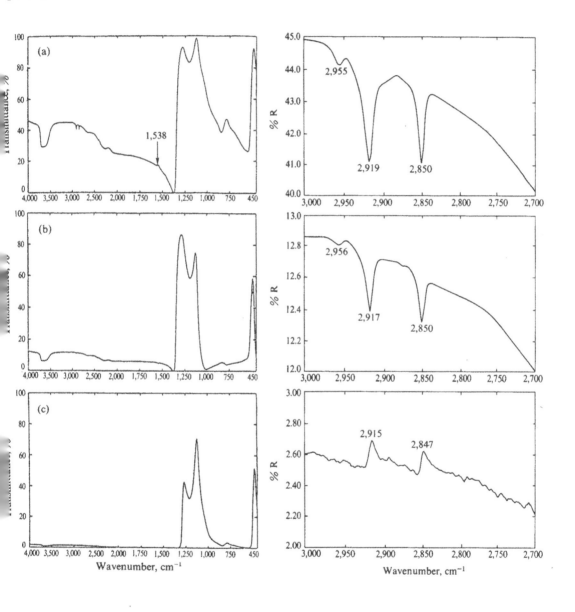

6.2E	LB Film Formed on Silica Plate (II)	External Reflection

This section deals with the spectral features of the IR-ER spectra obtained from silica plates coated with monolayer and nine-layer LB films.   Since the 1-mm thick silica plate is transparent to IR radiation in the region of 4,000 – 2,100 cm^{-1}, incident IR radiation passes through the LB film and silica and is partially reflected at the back side of the silica, which is also coated with the LB film, and the reflected radiation comes back to the front surface.   Therefore, the theoretical explanation of the CH stretching vibrations seems to be complicated due to the fact that detected IR radiation contains both reflection and transmission components of the LB film.   The experimental results of C–H stretching vibrations will be explained separately from the other vibrations, because the present example represents a special case of IR-ERS when the substrate is optically transparent.   In fact, the phase and relative intensities of CH stretching vibrations with s-polarized light do not agree with Tanaka's observations[1], which were performed on a sample without a transmission component.   On the other hand, all of these features agree well with Tanaka's description of the p-polarization experiment, in spite of the fact that there is a transmission component in the present case.   An explanation follows.

Difference spectra between IR-ER spectra taken from the silica with LB film and the reference silica were calculated in order to isolate the spectral features of the LB film from the silica spectrum.   Chart (a) and (b) display some of the difference spectra in absorbance units taken with p- and s-polarized light at 5° increments from 25° to 75° angle of incidence, respectively. As the incident angle of s-polarized light increases from 25° to 75°, the intensities of the C–H stretching vibrations changed smoothly, passing the maximum value at about 60°.   Note that the intensities are all positive in absorbance units, being contrary to Tanaka's case.   On the other hand, the C–H stretching vibrations for p-polarized light showed two discontinuities in phase and intensity.   Discontinuities occur at around 30° and 50° angles of incidence, the latter of which is close to the Brewster angle of the silica.   At the Brewster angle, the light changes phase 180° by a reflection; this is a reasonable cause for the observed phase change at 50°.   As explained above, angular dependence of external reflection spectra may show complications when the sample is not opaque.

1.   A. Udagawa, T. Matsui, and S. Tanaka, *Appl. Spectrosc.*, **40**, 794 (1986).

Chart 6.2E

<Experimental conditions>
4 cm^{-1} resolution, DTGS detector, 100 scans, SPECAC "Monolayer" variable angle specular reflection accessory with AgBr wire-grid polarizer

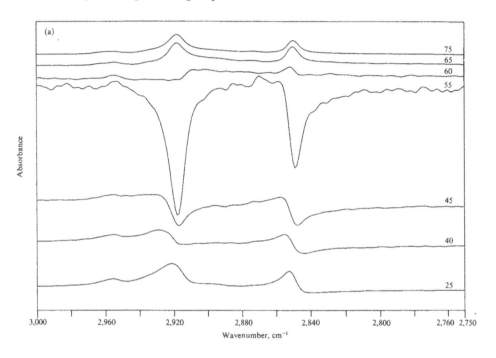

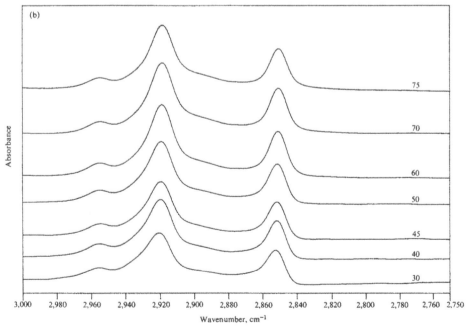

| 6.2F | LB Film Formed on Silica Plate (Ⅲ) | External Reflection |

In this section, the features of COO⁻ stretching vibrations and CH₂ bending and CH₂ wagging vibrations, all of which were measured in the frequency region where the silica is not optically transparent are discussed. This assumes that the experimental conditions satisfy the hypothetical experimental conditions of IR-ERS where only the IR radiation reflected at the front surface of the sample is detected.

A part of the IR-ER spectrum with a 70° angle of s-polarized light is shown in Fig. 6.5. In order to examine the spectral features in more detail, the features due to silica were removed by spectral subtraction. The peak at 1,560 cm⁻¹ assigned to $\nu_{as}(COO^-)$ and a doublet at 1,474 and 1,464 cm⁻¹ assigned to CH₂ bending mode of the crystalline CH₂ zigzag chain were found to show negative intensities. In addition, a peak at 1,424 cm⁻¹, which was assigned to $\nu_s(COO^-)$, and a band progression were observed at some incident angles as shown in Fig. 6.6. Chart (a) the optimum incident angle for the largest amplitude of the $\nu_{as}(COO^-)$ peak in IR-ERS was found to be around 30°. The intensity of the $\nu_{as}(COO^-)$ with a 30° angle of incidence with s-polarized light was six times as intense as that of a 75° angle of incidence.

Although measurements with the 55° angle of incidence with p-polarized light showed both $\nu_s(COO^-)$ and CH₂ bending vibration with *positive* absorbance, $\nu_s(COO^-)$ band with *negative* absorbance was found as shown in chart (b).

All of the $\nu_{as}(COO^-)$, $\nu_s(COO^-)$, and CH₂ bending bands in the case of s-polarized light showed negative absorbance regardless of the incident angle, while all of those bands measured with p-polarized light changed phase before and after the Brewster angle. It is unfortunate that the peaks due to $\nu_s(COO^-)$ at 40° and 45° angles of incidence are not clear due to the strong spectral features of silica below 1,400 cm⁻¹. In the case of observations with s-polarized light, an electric field of IR light lies in the y-direction in the surface layer. (refer to Fig. 2.22 for definition of x, y, and z axes.) Thus, the electric field interacts with vibrations whose induced dipole moment lies in the y-direction. In the case of incidence with p-polarized light, an electric field has two components, the z component, which is normal to the substrate, and the x component, which lies in a molecular layer and in the unclear direction of light propagation. Thus, p-polarized light

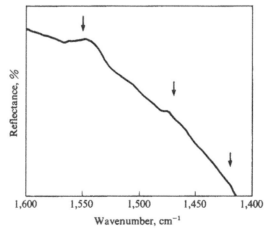

Fig. 6.5   IR-ER spectrum of glass substrate coated with ninelayers of cadmium arachidate.   LB film measured at 70° angle of s-polarized incidence.

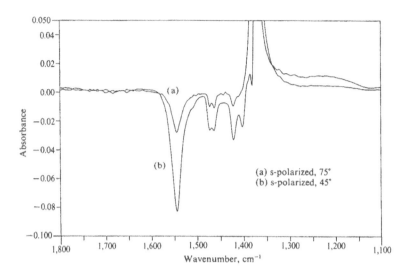

Fig. 6.6 Difference spectra showing spectral features of cadmium arachidate. LB film measured at 45° and 75° angle, of s-polarized incidence.

interacts with vibrations whose induced dipole moment is either normal to the substrate or lies in the molecular layers. (The reader may recall that the y and x components do not exist when the substrate is a *metal*.) Takenaka *et al.* showed the reflection absorbance for s-polarized light to be

$$A_s = -\frac{4}{\ln 10}\left(\frac{\cos\theta}{n_3^2 - 1}\right)\cdot n_2\cdot\alpha_{2y}\cdot h_2 \tag{6.1}$$

where n_2 and n_3 are the refractive indices of molecular layers and substrate, respectively. α_{2y} and h_2 are a y component of absorption of molecular layer and thickness of the layers, respectively. On the other hand, the x and y components of signal intensity in the case of observations with p-polarized light are given by

$$A_{px} = -\frac{4}{\ln 10}\left(\frac{\cos\theta}{\xi_3^2/n_3^4 - \cos^2\theta}\right)\cdot\left(-\frac{\xi_3^2}{n_3^4}\right)\cdot n_2\cdot\alpha_{2x}\cdot h_2 \tag{6.2}$$

$$A_{pz} = -\frac{4}{\ln 10}\left(\frac{\cos\theta}{\xi_3^2/n_3^4 - \cos^2\theta}\right)\cdot\left(\frac{\sin^2\theta}{(n_2^2 + k_{2z}^2)^2}\right)\cdot n_2\cdot\alpha_{2z}\cdot h_2 \tag{6.3}$$

where $\xi_3 = (n_3^2 - n_1^2\cdot\sin^2\theta)^{1/2}$.

Since the induced dipole moment of $\nu_{as}(COO^-)$ lies in the x- and y-directions, it is clear that this vibration shows a negative intensity for s-polarized light and positive intensity for p-polarized light and high incident angle, θ, so that the denominator of Eq. 6.2 is positive. On the other hand, $\nu_s(COO^-)$ at 1,430 cm^{-1} has an induced dipole moment in the z-direction. Therefore, as Eq. 6.3 indicates, the intensity of this peak should be negative when the incident angle is so large that the denominator is positive. However, when the incident angle becomes small enough to change the sign of the term $\xi_3^2/n_3^4 - \cos^2\theta$, to negative, absorbance becomes positive as observed.

1. T. Hasegawa, J. Umemura, and T. Takenaka, *J. Phys. Chem.*, **97**, 9009 (1993).

Chart 6.2F

<Experimental conditions>
4 cm^{-1} resolution, narrow band MCT detector, 300 scans coadded for the nine-layer sample and 3,000 scans coadded for the monolayer sample, SPECAC "Monolayer" variable angle reflection acccessory with AgBr wire-grid polarizer

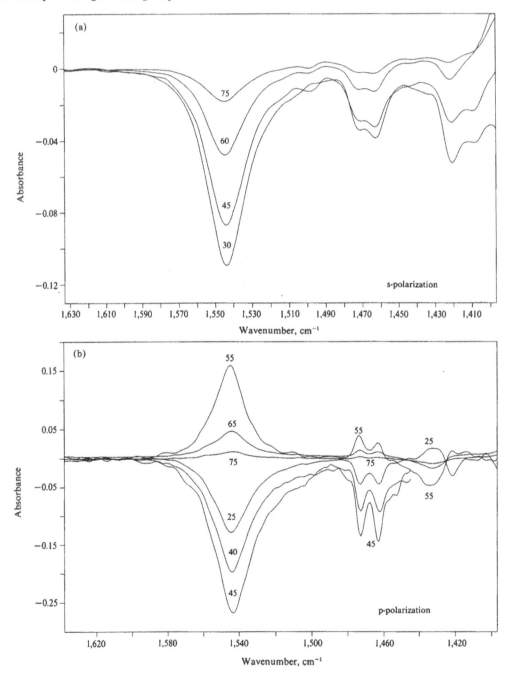

6.3 Thin Oxide Layers Formed on Metal Surfaces

6.3A	**Barrier Type Oxide on Alminum (I)**	Grazing Angle IR-RAS with KK Analysis

Thorough IR spectroscopic studies were performed on the barrier oxide layer on aluminum, since the barrier oxide is related to the anti-corrosive protection and coloring of aluminum products. According to Hannah's work[1] with IR-RAS, the barrier type oxide layer showed a peak at 960 cm^{-1} and it was assigned to alumina. Fig. 6.7 shows the IR-RA spectra Hannah observed in 1963 using a dispersive prism spectrometer from the 28 Å(A), 56 Å(B) and 84 Å(C) thick oxide layers. After careful debate,[2] the surface material has been assigned to γ-Al$_2$O$_3$ and the peak at 960 cm^{-1} is assigned to the Al–O–Al stretching vibration shifted from the 738 cm^{-1} peak position in the absorption spectrum. Moreover, the Al–O–Al bending vibration at 590 cm^{-1} was noted as shifting to 700 cm^{-1} in the IR-RA spectra. However, these large shifts were only intuitively explained in terms of the influence of the real part of the refractive index, n, of the surface material.[3]

Since grazing angle reflection-absorption spectra are theoretically formulated as given by equation (2.28), we calculated the grazing angle IR-RA spectrum of barrier type oxide on aluminum plate by KK analysis of the absorption spectrum of various kinds of alumina and related compounds, e.g. α-, κ-, χ, η-, and γ-alumina, aluminum oxide hydroxide, aluminum oxide trihydroxide, and Boehmite.[4] Among the candidates for surface species, only the low temperature transition alumina such as γ- and η-alumina gave an IR-RA spectrum which

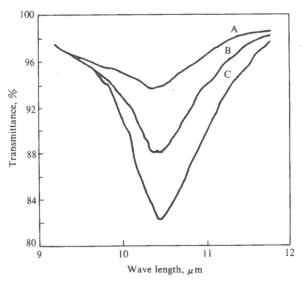

Fig. 6.7   Grazing angle IR-RA spectra of a barrier type oxide of aluminum plate.

1.   R.W. Hannah, *Appl. Spectrosc.*, **17**, 23 (1963).
2.   (a) G. A. Dorsey, Jr., *J. Electrochem. Soc.*, **113**, 284 (1966); (b) W. Vedder and D. A. Vermilyea, *Trans. Faraday Soc.*, **65**, 561 (1969); (c) F. P. Mertens, *Surface Sci.*, **71**, 161 (1978); (d) R. A. Ross and R. Lemay, *Surface Technol.*, **26**, 125 (1985).
3.   G. W. Poling, *J. Electrochem. Soc.*, **116**, 958 (1969).
4.   K. Nishikida and R. W. Hannah, *Appl. Spectrosc.*, **46**, 999 (1992).

Chart 6.3A

<Experimental conditions>
8 cm^{-1} resolution, DTGS detector, 16 scans for CsI pellet of γ-alumina

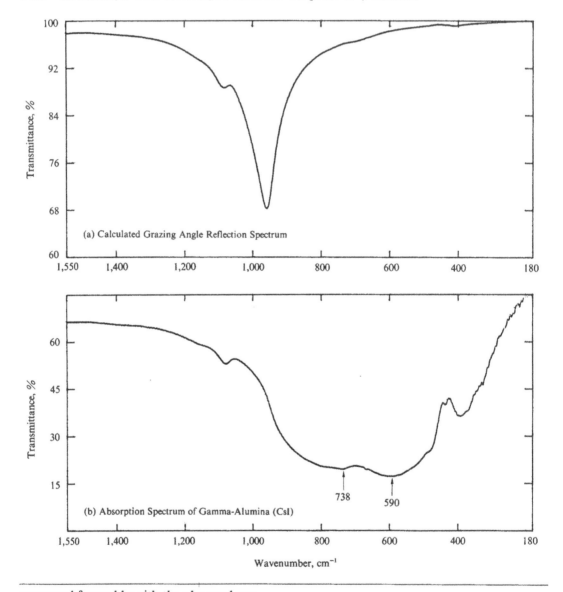

(a) Calculated Grazing Angle Reflection Spectrum

(b) Absorption Spectrum of Gamma-Alumina (CsI)

Wavenumber, cm^{-1}

compared fovorably with the observed one.

Spectrum (a) in the chart is the calculated spectrum, using the absorption spectrum (b) measured form a CsI pellet of γ-alumina to obtain the n and k values. In addition to a good coincidence in the peak position of the 970 cm^{-1} band with observed value at 960 cm^{-1}, a shoulder at *ca.* 690 cm^{-1} corresponds to the peak at 700 cm^{-1}.

It may be concluded from this that KK analysis is essential if the IR-RA spectra of thin surface layers on metals is to be understood.

6.3B	Barrier Type Oxide on Alminum (II)	Emission

Although the grazing angle IR-RAS is commonly employed to study thin oxide layers on metal substrates as discussed in the preceding section, it is possible to study similar samples by emission spectroscopy.   As explained in Section 2.6, use of the sample space emission accessory allows simple acquisition of emission spectra, even if there is no dedicated emission sample compartment in the analyst's FT-IR system.   In this section an experimental procedure to observe emission spectra of the barrier type thin oxide layer on aluminum plate is described.

A series of different thicknesses of the oxide layer on aluminum plates was prepared by moderate air oxidation in a hot oven, the thicknesses being estimated as being from 40 – 200 Å. An aluminum plate washed with diluted nitric acid was dried without heating and used as the reference aluminum plate.   None of the aluminum plates showed any (normal incidence) IR-RA spectra in the frequency region of 1,200 – 400 cm^{-1}.   However, with the grazing angle IR-RAS, all of the oxidized aluminum plates showed a peak at 960 cm^{-1}.   These two observations confirm that the surface layer is truly thin compared with the wavelength of the IR radiation, and the species is assigned to γ-alumina or the barrier type oxide layer.

The reference, or sample aluminum plate, was mounted in the furnace assembly of a goniometer type emission accessory and the emission spectra were measured with a TGS detector. The emissivity was calculated according Eq. 2.31.

Chart (a) illustrates the emission spectra measured at different observation angles at 200°C.* A peak due to the oxide layer was observed at 960 cm^{-1} in the emission spectrum and the peak intensity was larger with higher observation angles.   It is striking that the position of the oxide layer peak in the grazing angle emission spectrum is exactly the same as that of the grazing angle IR-RA spectrum.   Fig. 6.8 shows the correlation between the intensities of the grazing angle IR-RA spectra and the grazing angle emission spectra of the aluminum plate samples coated with thin oxide layer, a linear relationship suggesting that the interaction between the electric field of

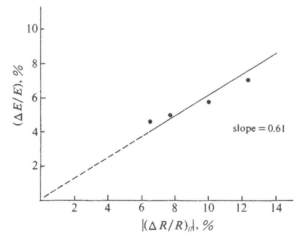

Fig. 6.8   Correlation between IR-RAS and emission intensities of a barrier type oxide layer of an aluminum plate.

* It was confirmed that none of the samples or reference aluminum plates was oxidized under a few minutes of heating at 200°C.

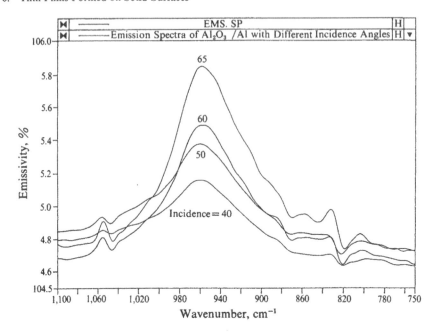

Fig. 6.9   Angular dependence of emission intensities of a barrier type oxide on an aluminum plate.

the emitted radiation and thin film in grazing angle emission spectroscopy is the same as that for the incident radiation in the grazing angle IR-RAS.   Fig. 6.8 also suggests that the emission spectroscopy introduced in this section can be utilized for quantitative analysis.

Let us discuss this phenomenon.   Since the emission radiation will be reflected at the metal surface, there is an interaction between the direct emission and reflected emission.   Being similar to the case of the grazing angle IR-RAS for thin layers on metal, this feature in the emission spectrum will be strongest when the emission is observed at the grazing angle shown in Fig. 6.9(a). On the other hand, there will be no feature in emission radiation if the sample is oriented normal to the interferometer and detector (Fig. 6.9(b)).   In addition, the peak position has a large frequency shift similar to the high frequency shift observed in the grazing angle IR-RAS.

Chart 6.3B

<Experimenta conditions>
8 cm^{-1} resolution, DTGS detector, 100 scans, Connecticut Instrument Company sample space emission accessory with goniometer and furnace

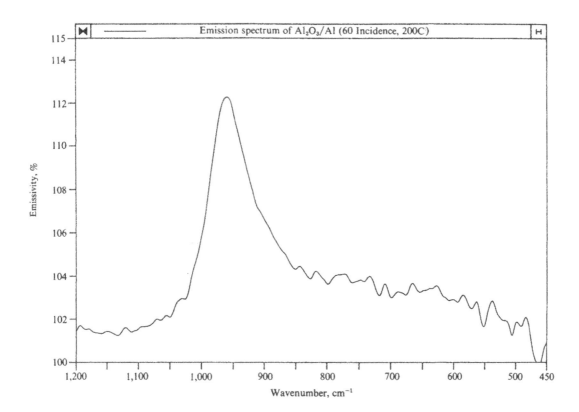

6.3C	**Barrier Type Oxide on Alminum (III)**	**ATR Method**

The samples used in the two previous sections were examined by the ATR technique.   Chart (a) shows the ATR spectrum measured with KRS-5 ($n = 2.37$) IRE with a 45° angle of incidence. The background spectrum in this case is obtained with the nitric acid-treated plate mounted in the ATR accessory.   The peak position due to the oxide layer, γ-Al_2O_3, is found to be 960 cm^{-1} and is the same as that observed by grazing angle IR-RAS and grazing angle emission spectroscopy. This observation indicates the same frequency shift from that of the 738 cm^{-1} band in the absorption spectra of γ-Al_2O_3, as noted previously.   Since the penetration depths of the evanescent waves at 1,000 cm^{-1} for the two ATR experiments, $ca.$ 2 μm for KRS-5 and 0.7 μm for Ge, are far larger than the thickness of the oxide layer (200 Å maximum), the ATR condition must be examined in terms of a three-body system, that is, KRS-5 (or Ge)/γ-Al_2O_3/Al.   In the ultimate situation when the thickness of the oxide layer is 0, the experimental conditions become the internal reflection at KRS-5/Al and Ge/Al boundaries with a 45° angle of incidence.   Since the refractive indices of KRS-5, Ge, and Al are 2.37, 4.0, and 1.44, the 45° angle of incidence is apparently larger than the critical angle of the KRS-5/Al system.   Although the experiment using a commercial ATR accessory with KRS-5 IRE at a 45° angle of incidence yields an ATR spectrum, the x and y components of standing wave are cancelled for the same reason that explains reflection of light at the surface of metal (see Section 2.4).   The z component of the incident radiation is reflected at the metal surface, creating standing wave at the metal surface equivalent to the standing wave of the grazing angle IR-RAS.   Therefore, the experimental condition employed here is the grazing angle IR-RA spectrum but the incident radiation is not coming from air but from the KRS-5 crystal.   In order to confirm that the present spectrum is a grazing angle IR-RAS and not an ATR spectrum, theoretical calculation of the spectrum based on a stratified layer model was carried out.   In this model, incident radiation from KRS-5 arriving at 100 Å thick Al_2O_3 formed on aluminum plate with infinitive thickness with 45° angle of incidence was considered.[1]   Charts (b) and (c) show the simulated spectra of the present experimental setting for s- and p-polarized light, respectively.   Simulation spectra indicate that the IR spectrum obtained is a grazing angle IR-RA spectrum.

As shown clearly in this section, the analyst must be cautious when employing reflection techniques to observe spectra of thin layers on substrate.   In Section 9.1G, the reader will find a different application of the ATR technique to enhance signal intensity when the surface material is very thin.

1.   K. Nishikida, Abstract #790, Pittsburgh Conference (1993) and Abstract #257, Pittsburgh Conference (1994).

Chart 6.3C

<Experimental conditions>
4 cm^{-1} resolution, 36 scans coadded, DTGS detector, Wilks type ATR accessory

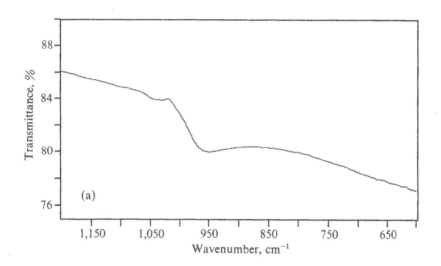

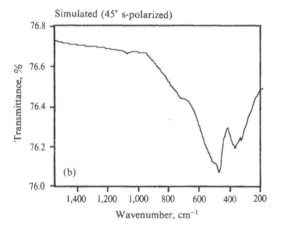

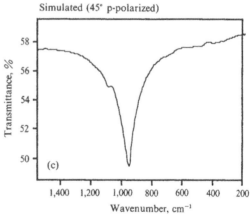

6.3D	Lubricant of Low Density Hard Disk Media	IR-RAS

The traditional hard disk media used for Winchester-type hard disks is an aluminum disk coated with an epoxide resin containing γ-Fe_2O_3.   The thickness of the coating material at inner radius is about 7,500 Å (0.75 μm) and the thickness gradually increases to 15,000 Å (1.5 μm) at the outermost radius.   The epoxide layer, which is the magnetic layer, is coated with a thin film of fluorocarbon-type lubricant to protect the magnetic layer from touch-down damage due to a reading device.   The thickness of the lubricant is one of the subjects of quality control in the production of hard disks using graing angle IR-RAS technique.   Although the thickness, d (0.75 – 1.5 μm), of the magnetic layer is small compared with the wavelength of the IR radiation (2.5 – 25 μm), the question is whether the thickness is thin enough to categorize the experiment as grazing angle IR-RAS, where the theoretical requirement is that the thickness is *very* small compared with the wavelength ($d \ll \lambda$).[1]

Charts (a) and (b) display the grazing angle and normal incidence IR-RA spectra (in %R units) of coated and uncoated hard disk media together with difference spectra (in absorbance units), respectively.   Comparing these spectra, it may be noted that the intensities of all of the bands, especially those in the low frequency region such as 830 and 710 cm^{-1} bands due to epoxide and the 670 cm^{-1} band due to γ-Fe_2O_3, are larger in the case of grazing angle IR-RAS.   In addition, the difference spectrum of the grazing angle IR-RAS is twice that observed at normal incidence IR-RAS, showing the advantage of employing the grazing angle arrangement.

Although the peak positions of bands observed in the IR-RAS arrangement due to epoxide coincide well with those of absorption spectrum,[2] the coincidence between peak positions in chart (a) and the absorption spectrum is not necessarily good.   The peak due to γ-Fe_2O_3 indeed shows a substantial frequency shift.[3]   However, some peaks due to the epoxide also show frequency shifts, although these shifts are not as large as for the γ-Fe_2O_3 bands.   Fig. 6.10 shows the position of the intense peak at 1,240 cm^{-1} due to the epoxide resin as a function of thickness. When the thickness is large (2.5 μm), the peak position approaches that of the absorption spectrum, while it shows a rather distinct high frequency shift when the thickness becomes smaller.

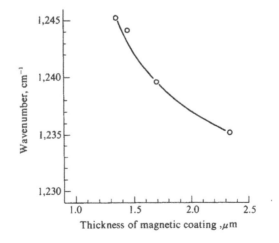

Fig. 6.10   Peak position of the epoxide resin absorption *versus* thickness of the epoxide resin.

Chart 6.3D

<Experimental conditions>
4 cm^{-1} resolution, DTGS detector, 16 scans, 80° angle of incidence for chart (a), 6.5° angle of incidence for chart (b)

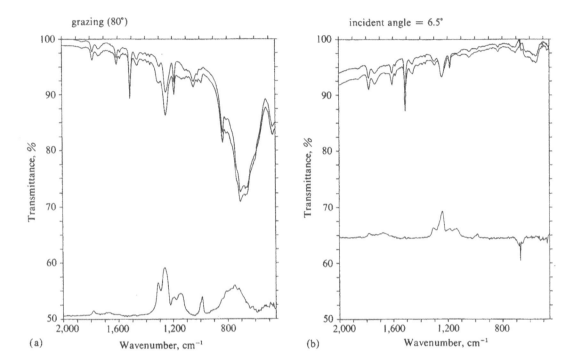

(a)

(b)

(1.3 μm).[4]   Thus, the thinner film shows a larger contribution due to the interaction with the surface electric field.   This indicates that the ratio of the contributions due to the usual absorption and that obtained for the portions of the film nearest to the metal surface varies as the thickness varies.   It should be also noted that the employment of the grazing angle benefits the non-boundary spectrum by a clear increase in the effective film thickness at the higher angle of incidence.

The lubricant spectra shown in charts (a) and (b) were compared.   Apart from the intensity differences, the line shapes are also different.   As in the case of the effect of the surface electric field on the epoxide spectrum, it is reasonable to conclude that the different band shapes are also due to the influence of the surface electric field, the result being expressed in terms of the product of n and k given in Eq. 2.28.   As discussed above, the 0.75 – 1.5 μm thick surface coating is too thick to interpret the spectrum in terms of the grazing angle IR-RAS.

1.   J. D. E. McIntyre, in: *Advances in Electrochemistry and Electrochemical Engineering,* vol. 9, p. 61, John Wiley & Sons (1973).
2.   Chicago Society for Coatings Technology ed., *An Infrared Spectroscopy Atlas for the Coating Industry,* Federation of Societies for Coatings Technology (1980).
3.   G. W. Poling, *J. Electrochem. Soc.,* **116**, 958 (1969).
4.   Therefore, it is essential to measure the matched positions of the hard disk media when the difference spectroscopy is intended to quantitize the thickness of the lubricant.

6.3E	Lubricant of High Density Hard Disk Media (I)	Grazing Angle IR-RAS

Most hard disk media used in personal and laptop computers are of the high density magnetic disk type.   According to a review article, an aluminum or aluminum alloy disk is sputter-coated with a nonmagnetic alloy and CoNiP magnetic alloy (*ca.* 80 Å), the latter being the top layer. The magnetic layer is protected with a thin (*ca.* 300 Å) amorphous carbon film.   Further, a thin (10 – 30 Å) fluorocarbon type lubricant is applied on the amorphous carbon.   Since CoNiP reflects the IR radiation and the dielectric layers (amorphous-carbon and lubricant) on CoNiP are both thin enough to fulfill the criteria for the interaction between the surface electric field and the substance expressed within the theory of the grazing angle IR-RAS, the high density hard disk media seem to be an ideal material to examine the characteristics of grazing angle IR-RAS.

In theory, the interaction between IR radiation and thin materials on metal plates arises when p-polarized radiation is introduced at a grazing angle, *i.e.*, near 80°.   Four experiments have been performed to demonstrate the phenomenon.

Chart (a) shows that the intensity of a peak due to the fluorocarbon lubricant becomes higher as the incident angle becomes higher, p-polarized radiation being used.   According to Eq. 2.28, the incident angular dependence of the peak intensity is expressed as $I = \sin^2 \theta / \cos \theta$ (θ: incident angle).   With $\theta = 60°, 70°$, and $80°$, Eq. 2.28 predicts the intensity to be $1 : 1.7 : 3.7$, respectively, while the observed values are $1 : 1.5 : 3.1$, respectively.

Chart (b) demonstrates that the thin film does not show any features at a normal angle of incidence, while the grazing angle incidence gives rise to an enhanced feature in reflection spectrum with p-polarized radiation.

Chart (c) shows that p-polarized radiation has features at the 80° incident angle, while s-polarized radiation does not.   This confirms the theoretical expectation, as explained in Section 2.4, that even with high incident angle s-polarized radiation does not give rise to any spectral feature of thin film in reflection spectra.

Chart (d) proves that the use of the IR polarizer for p-polarization enhances the intenstiy of thin film features.   It was found from chart (d) that the intensity with unpolarized radiation dropped to 1/2.7 that with p-polarized radiation.   Assuming that the intensities of p- and s-polarized radiations are equal and the metal reflectance of p- and s-polarized radiation are $R_{/\!/}$ and $R_\perp$, respectively, although s-polarized radiation provides zero change in the reflectivity, p-polarized radiation changes reflectance by ΔR due to the thin film.   Thus, the relative reflectance change in the case of unpolarized is given by $\Delta R / (R_{/\!/} + R_\perp)$, while that of p-polarized radiation become ($\Delta R / R_{/\!/}$).   Provided that $R_{/\!/} = R_\perp$, the intensity of feature in the reflection spectrum in the case of unplarized radiation becomes one-half that in the case of p-polarized radiation.   Usually $R_\perp > R_{/\!/}$, which means that the observed intensity without a polarizer is less than half that obtained with a polarizer.

Thus, these four charts confirm every aspect of the McIntyre-Suetaka theoretical predictions for grazing angle IR-RAS.

Chart 6.3E

<Experimental conditions>
4 cm^{-1} resolution, DTGS detector, 20 scans, variable angle specular reflection accessory and AgBr-Gold wire grid polarizer

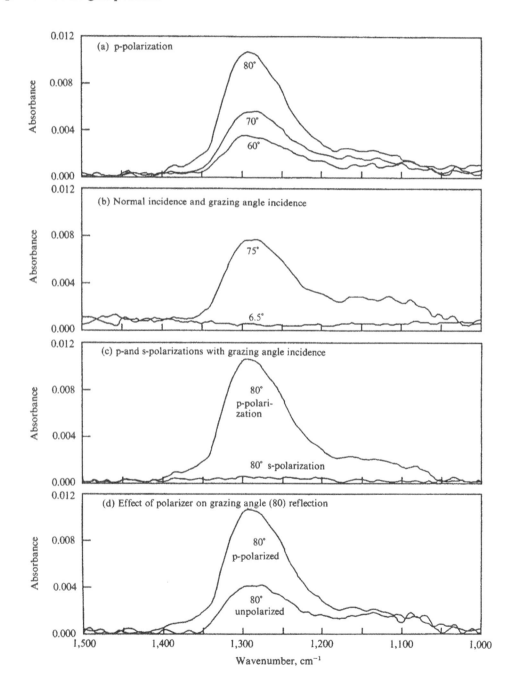

6.3F	Lubricant of High Density Hard Disk Media (II)	Grazing Angle IR-RAS

Since the grazing angle IR-RAS was found to have adequate sensitivity to quantitize the thickness of lubricant of high density hard disk media, an unused and used 5-inch high density hard disk media were examined to determine the distribution of the lubricant over the entire area, using a dedicated automated hard disk checker.[1]   Unfortunately, it is not possible by the grazing angle IR-RAS to directly calibrate the system to determine the thickness from the peak intensity without standard materials.   Thus, a set of six hard disk media for which the thickness was determined by ESCA (Electron Spectrosocopy for Chemical Analysis)[2] was used to calibrate the system.   Measurements of the IR spectra were made every 3 mm in diameter starting from 18 mm and going to 42 mm, and every 15° around the disk.   Since measurement of all 216 positions, 9 linear and 24 angular, takes about 16 hours, the time stability of the system was first checked.   Fig. 6.11 shows the peak intensity of the lubricant measured every 20 minutes for 64 hours, using only one background spectrum measured at the start of the experiment.   The relative standard deviation of the observed values was 1.6% for the average value of 0.091 absorbance at the peak.   This result ensures that the system has enough stability to make a detailed study of lubricant thickness as a function of position.

In the charts, 3D displays of the lubricant thickness over the (a) unused (experimental) and (b) used (production) high density hard disk media are shown.   The lubricant was almost evenly applied over the entire area when it was manufactured as chart (a) shows.   However, after a prolonged quality assurance test, in which the medium was rotated under the same ambient condition as in actual use in the Winchester hard disk drive, it was found that the lubricant migrates from the inner to the outer area, as clearly illustrated in chart (b).

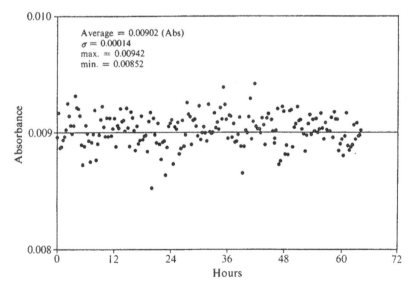

Fig. 6.11   Peak intensities (in absorbance units) of lubricant peak measured from the same position every 20 minutes for 64 hours, using the same background spectrum measured just before the first measurement at time 0.

1.   K. Nishikida, Abstract #779, Pittsburgh Conferences (1992).
2.   Courtesy of Alps Electronic Corporation, Nagaoka, Japan.

Chart 6.3F

<Experimental conditions>
8 cm^{-1} resolution, DTGS detector, ambient temperature controlled within ±1°C, 20 scans, Perkin-Elmer automated disk checker equipped with KRS-5/Al wire grid polarizer

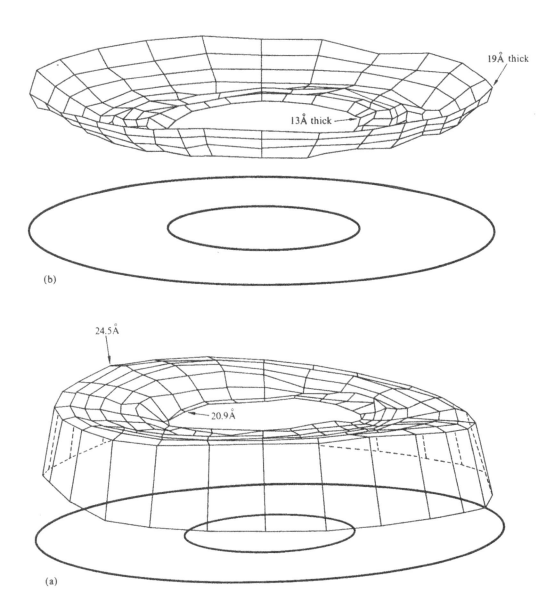

(b)

(a)

7. Liquid Samples

7A	Association of Glycin Oligomer	ATR

It is well known that the crystal structure of polyglycin, which is a polypeptide composed of only one kind of amino acid, glycin, is either a β-pleat structure (type-I) or an α-helix structure (type-II).[1]   Low molecular weight oligomers from, for example, dimers to quintuple oligomers have a β-pleat structure while oligomers with 6 to 12 monomer units have an α-helix structure. In this section, the change in the higher order structure of N-butanoyl–(tri and tetra)–glycylglycin potassium salts due to association in water solvent as studied by ATR technique is presented as an example of ATR application to diluted water solutions.

Charts (a) and (b) show the ATR spectra of the tetramer and trimer, respectively.   ATR (ZnSe/45°)[2] spectra of the solutions were measured and features due to water were cancelled from all spectra using computer software.   These charts demonstrate the spectral changes which occur as a function of concentration of oligomers.   Two peaks at 1,030 and 1,018 cm^{-1} observed in low concentration such as 1.4×10^{-2} M in chart (a) represent the second-order structure of these oligomers.   The former peak is assigned to the α-helix and the latter to the β-pleat structure.   In Fig. 7.1 the ratio of absorbance at 1,030 cm^{-1} to that at 1,018 cm^{-1} was plotted[3] *versus* concentration.   A distinct point is shown in Fig. 7.1.   The concentration at this point

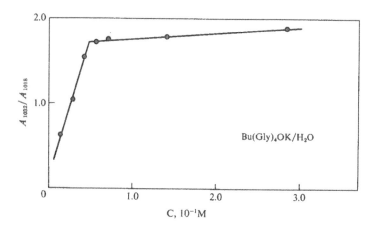

Fig. 7.1   Dependence of ratio between I-form and II-form on concentration.

---

[1]   Refer to Section 8C for higher structure of polypeptides.
[2]   KRS-5 IRE should not be used for aqueous solutions because the IRE slightly dissolves in water.   As a result, the optical throughput largely deteriorates and the IRE cannot be reused unless it is repolished by professionals.
[3]   Although the 1,018 cm^{-1} peak is not seen in the spectra at high concentrations, deconvolution techniques followed by a curve fitting enables resolution of the overlapping band into two components.

corresponds exactly to the critical micelle concentration (CMC) of the oligomer.   Below the CMC, the ratio increases linearly with concentration.   This indicates that the oligomer takes an open pleat structure when the concentration is low, and as the concentration increases the content of the helix structure increases.   At concentrations higher than the CMC, the change in the ratio of helix to pleat structure is not large, reflecting the fact that the oligomer takes on a uniform structure in the micelle at higher concentrations.   The spectra taken from the same experiment on the potassium salt of N-butyl–glycyl–glycyl–glycin are shown in chart (b).   It was found that the trimer also assumes a pleat structure at low concentrations, and that the helix form increases with increase in the concentration until the concentration reaches the CMC of the trimer.   Also, it is shown that the molecular structure becomes uniform at concentrations above the CMC.

1.   E. Nishio, H. Natori, S. Sugiyama, K. Taga and H. Okabayashi, Proceedings of the Forty Colloid and Sufactant Science, p. 318, (1987).

Chart 7A

<Experimental conditions>
4 cm^{-1} resolution, DTGS detector, 16 scans ATR accessory: versatile type vertical MIR with solution accessory equipped with ZnSe IRE at 45° angle of incidence

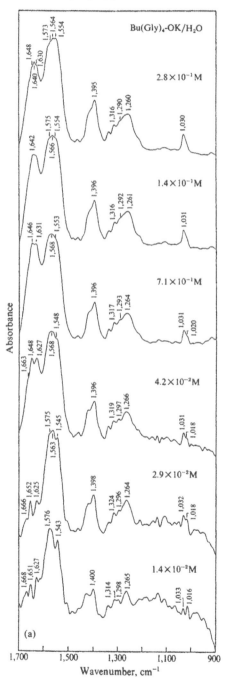

(a)

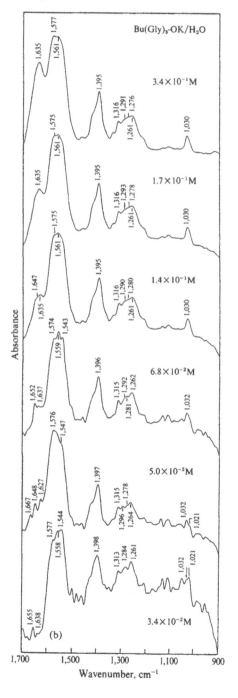

(b)

7B	**Association of Surfactant in Water**	ATR

Using the same experimental procedure described in the preceding section, the higher order structure in water solution of potassium *trans*-4-hexenoate (PT4H), whose chemical structure is that of a simple surfactant, was examined by ATR technique.[1]

$$CH_3-CH=CH-CH_2-CH_2-COOK$$
$$(\omega \quad \delta \quad \gamma \quad \beta \quad \alpha)$$

The possible isomers of the T4H ion are TS, TC, GS, G′S, and GC (T: *Trans*, C: *cis*, G: Gauche, S: skew) among all these possible structural isomers composed of combinations between R′–CH_2–CH_2–R (R′=COO^- and R=–CH=CH–) and R′–CH_2–CH=CH–R (R′=–CH_2– and R=CH_3).[1] In the frequency region between 1,350 – 1,000 cm^{-1} of the IR spectra of PT4H shown in Chart (a), some peaks corresponding to these rotational isomers have been assigned.   For instance, peaks at 1,236 and 1,263 cm^{-1} are assigned to a twisting vibration of the –$C^\alpha H_2$– group in the TS isomer and an in-plane bending vibration of the –CH=CH– of isomers other than the TS isomer, respectively.[1]   The absorption intensities of the 1,263 and 1,236 cm^{-1} peaks are plotted against the reciprocal of the concentrations, $1/c$, as shown in Fig. 7.2.   It is clear that both absorption intensities abruptly change at a unique concentration corresponding to the CMC of PT4H.   As the concentration of PT4H increases, the intensity of the 1,236 cm^{-1} band increases at the expense of the 1,263 cm^{-1} band.   In addition, the slope changes at concentrations higher than the CMC, revealing that the TS structure becomes more stabilized with increase in concentration, and the structure changes abruptly when the micelle is formed.

Although some absorption bands are known below 1,000 cm^{-1}, it is not possible to observe these bands in aqueous solution, since the strong absorption due to water prevents IR observation

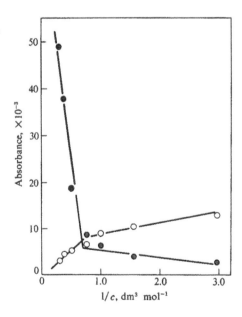

Fig. 7.2   Dependence of peak intensities on concentration of PT4H.   (o; 1263 cm^{-1}; •; 1236 $^{-1}$)

Chart 7B

<Experimental conditions>
4 cm^{-1} resolution, DTGS detector, 16 scans

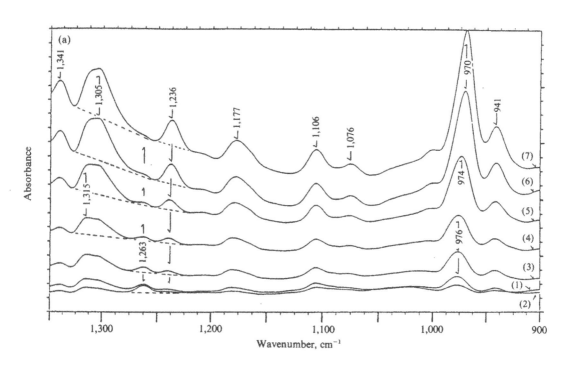

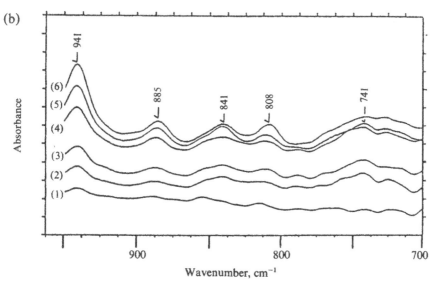

in this region.   However, use of deuterated water (D$_2$O) as a solvent enables observation of IR spectra in the region of 1,000 to 700 cm^{-1}.

Chart (b) demonstrates the usefulness of D$_2$O as a solvent in aqueous solution studies to incorporate this lower frequency region.   Peaks at 808 and 841 cm^{-1} are assigned to a rocking vibration of –CH$_2$CH$_2$– of the TS isomer and of TC and GS isomers both, respectively.   Peaks at 770, 756, 741, 725, and 715 cm^{-1} are assigned to an out-of-plane bending vibration of the –CH=CH– group of GS, GS′, TS, GC, and TC isomers, resepectively.   Spectra in chart (b) show the increase in the intensities of both the 808 and 741 cm^{-1} peaks of the TS isomer to increase with concentration, corresponding to an increase in the TS isomer content.   These observations confirm the conclusion deduced from the observation in the 1,350 to 900 cm^{-1} region.

1.   H. Okabayashi, K. Tsukamoto, K. Ohshima, K. Taga and E. Nishio, *J. Chem. Soc., Faraday Trans. 1*, **85**, 1639 (1988).

7C | Difference Spectroscopy and Molecular Interaction

One of the most powerful applications of FT-IR spectroscopy is computer-based difference spectroscopy. Difference spectroscopy is a computational procedure to calculate the spectrum of a pure component from the spectra taken from different compositions, using a primitive mathematical technique, substraction of two spectra. Let us suppose the absorption spectra (in absorbance units) of components A and B as $A(\tilde{\nu})$ and $B(\tilde{\nu})$, respectively. When the fractions of A and B are a and b, respectively, the IR spectrum of a mixture of compounds A and B is written as

$$M(\tilde{\nu}) = a \cdot (\tilde{\nu}) + b \cdot B(\tilde{\nu})$$

When another spectrum, $M'(\tilde{\nu})$, of the mixture with different fractions, a' and b', is given, a difference spectrum, a spectrum of component A for example, is obtained by computing the following

$$M(\tilde{\nu}) - (b/b') \cdot (\tilde{\nu}) = c \cdot A(\tilde{\nu})$$

Difference spectroscopy is frequently utilized in modern IR spectroscopy and it is worth discussing a few applications in this volume.

However, it should be noted that a perfect cancellation of one of the components will be achieved only when the IR spectrum of each component in the mixture is the same as that of the pure material. Since the property of a compound changes with the environment, its IR spectrum will be influenced. Although the formation of hydrogen-bonds[1] or charge-transfer complexes[2] causes distinct changes in the IR spectra of the components, weak molecular interactions between components produce very small, often visually unobservable, spectral changes in the spectrum.

Chart (a) is the observed IR spectrum of a mixture of o-, m-, and p-xylene. Chart (b) is a computer-composed spectrum of the mixture, in which the IR spectra of pure components were

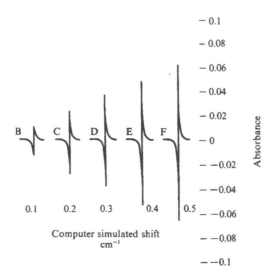

Fig. 7.3   Relationship between amplitude of the difference spectrum and shift.

added to the memory of the computer to match the composition of the real sample.

Chart (c) shows the difference spectrum, spectrum (a) – spectrum (b).   Although a doublet band at *ca*. 840 cm^{-1} is completely subtracted, other peaks except the one at the lowest frequency (marked with *M*) are left as derivative type lines.   The derivative type line in the difference spectrum appears when the sample line position is shifted from the reference position, the shift being due most probably to the molecular interaction among the components in the mixture.

Figure 7.3 shows the amplitudes of the derivative lines produced when difference spectra are calculated from the observed lines shifted in the memory of the computer by 0.1 to 0.5 cm^{-1} at every 0.1 cm^{-1} interval, and the unshifted line.   As the shift becomes bigger, the amplitude of the derivative line becomes bigger.   It is possible to estimate the shift from the amplitude of the derivative line, if the proper calibration is performed.

1.   (a) J. P. Leickman, J. Lascombe, N. Fuson and M. L. Josien, *Bull. Soc. Chim. Fr.,* 1516 **1959**; (b) G. C. Pimentel and A. L. McClellan, *The Hydrogen Bond,* p. 91, Freeman, San Francisco (1960).
2.   Y. Matsunaga, *J. Chem. Phys.,* **41**, 1609 (1964); *ibid.,* **42**, 1982 (1965).

Chart 7C

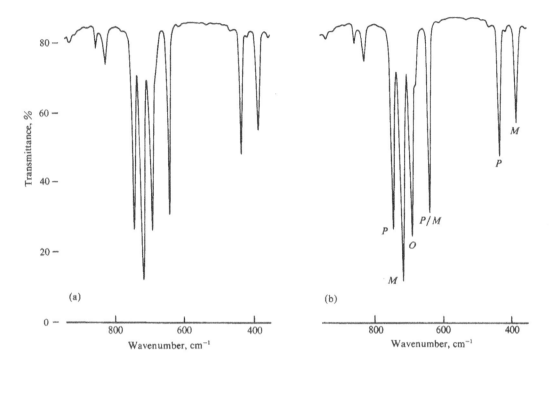

(a)

(b)

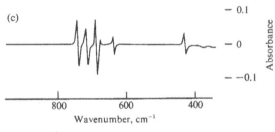

(c)

7D	Interaction between Polymer and Organic Acid in Solution[1]	Transmission

A polymer which is capable of bonding with organic acids could be used as adsorbing medium to improve the efficiency and specificity of acid recovery from solvents.   In order to evaluate the interaction between the organic acid and polymer, the IR spectra of a homogeneous polymer solution containing an organic acid was examined.   A copolymer of styrene and 4-vinylpyridine, VP–ST, was used as the adsorbing polymer and palmitic acid ($CH_3–(CH_2)_{14}–COOH$) was selected as the organic acid.   Perchloroethylene (PCE) was used as the solvent.

Spectra (a), (b), and (c) in the chart show the IR spectra of the three-component solution, the VP–ST film, and PCE, respectively.   Spectrum (d) is a difference spectrum $(a-(b+c))$ representing the palmitic acid in the three-component solution.   As the reference, an IR spectrum of the palmitic acid cast film on a KBr plate is presented in spectrum (e) of the chart.   In Fig. 7.4, the intensities of the CH_2 anti-symmetric stretching vibration at $2,925\ cm^{-1}$ and the intensity of the carbonyl stretching vibration in the dimer form at $1,710\ cm^{-1}$ were plotted against concentrations of the acid for those cases where VP–ST is present (solid circles and triangles) and absent (open circles and triangles).   The intensity of the carbonyl group relative to the CH_2 group becomes smaller when VP–ST is added to the solution, compared with that when the VP–ST is not present.   This implies that the fraction of the dimer form of palmitic acid decreased due to interaction with VP–ST.   Since the lone-pair electrons of the pyridine ring in the VP–ST acts as a base, interaction between VP–ST and the acid would most probably happen at the nitrogen atom in the pyridine ring.   So far efforts to find an absorption due to a $–CO_2^-$ group have been

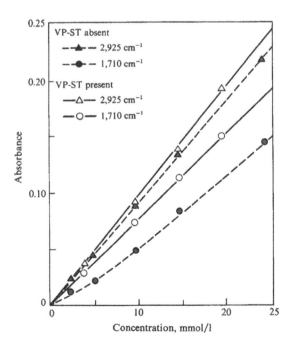

Fig. 7.4   Intensity of 2,925 and 1,710 cm⁻¹ peaks as a function of palmitic acid concentration.

Chart 7D

<Experimental conditions>
4 cm^{-1} resolution, DTGS detector, 60 scans

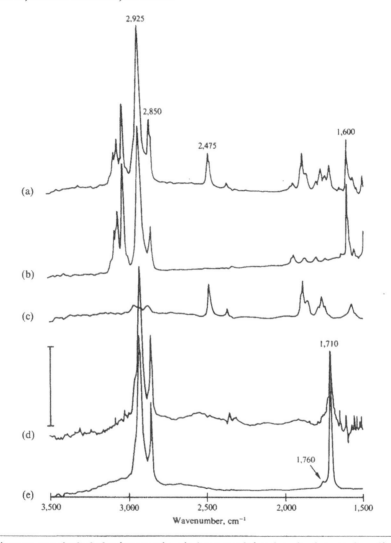

in vain, and it was concluded the interaction is best explained as hydrogen bonding as written below:

A weak band seen at 1,760 cm^{-1} in spectrum (e) is assigned to the carbonyl group stretching vibration in the monomer form of palmitic acid.

1. E. Nishio and N. Ikuta, Abstracts, 20th Conference of Applied Spectrometery (1984), Tokyo.

| 7E | **Analysis of Milk** | **ATR** |

Determinations of fresh milk components, *i.e.*, fat, protein, and sugar, by classical methods are already established.   Different methods are applied for each component and each method requires its own unique pretreatment of the milk sample.   However, it has been reported[1] that an IR spectroscopic method can determine all of these components from an IR spectrum.   Since more than 90% of the milk is water, transmission using a thin liquid cell, or the ATR technique seems convenient.   The use of ZnSe or AgCl cells with 15-μm path lengths are required for transmission work to avoid total absorption of IR radiation by the water.   On the other hand, it seems more convenient to use an ATR accessory with a cylindrical IRE, or the horizontal type of ATR unit, since filling and cleaning an ATR accessory is far easier than cleaning and refilling a thin liquid transmission cell.

In chart (a), the IR spectra of two different kinds of fresh milk are presented together with that of distilled water as a reference.   Chart (b) is the difference spectrum between one of the milk samples (upper) and water.   The determination of the fat component can be achieved by calculating the intensity of the CH stretching vibration at 2,850 cm^{-1} and the carbonyl stretching vibration at 1,740 cm^{-1}.   Protein can be determined from the intensity of a band at 1,540 cm^{-1}, the amide-II band, since the NH band at around 3,400 cm^{-1} is not available due to the strong interference from water.   The band at 1,650 cm^{-1}, which is the amide-I band, should not be used, since a strong absorption band of water overlaps this band also.

The content of sugar can be determined from the characteristic bands at around 1,100 – 1,000 cm^{-1}, assigned to the bending vibration of the OH group in sugars.

It is essential to homogenize the milk to avoid quantitation errors associated with heterogeniety of fresh milk.

1. Tetsuo Satoh, *Food Chemicals*, No. 9, (1987).

Chart 7E

<Experimental conditions>
4 cm⁻¹ resolution, DTGS detector, 53 scans
ATR accessory: Spectra-Tech Circle-Cell

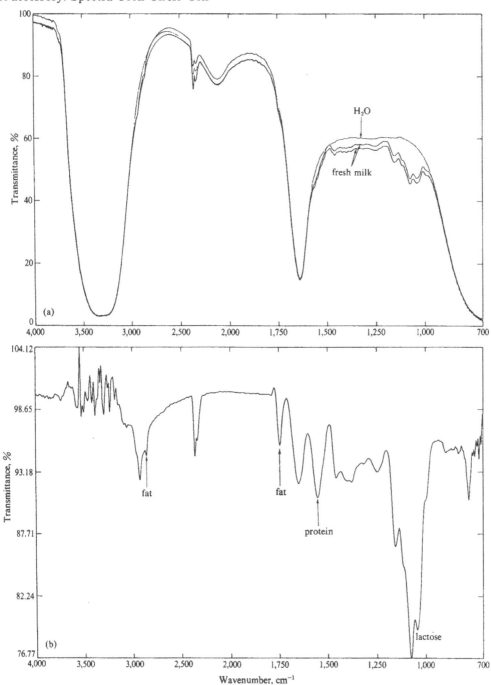

8. Biological Samples

8A	IR Spectrum of Bacteria	Transmission

Biological samples such as bacteria show a characteristic IR spectrum, Fig. 8.1, in which three peaks at *ca.* 3,300, 1,650, and 1,550 cm^{-1} are unambiguously assigned to the NH stretching vibration, amide-I, and amide-II bands of secondary polyamide (protein), respectively. In addition, a broad band with two or three peaks at 1,140 – 1,050 cm^{-1} is due to polysaccharides, and the band near 1,250 cm^{-1} is assigned to PO$_2^-$ groups. Since the possibility of identifying biological samples by IR spectroscopy was discussed early in 1911,[1] some studies were made to catogorize the bacteria.[2] A modern technique to categorize bacteria by IR spectra utilizes an FT-IR system and computer software. As Naumann showed,[3] the technique utilizes the first to fourth derivative spectra of the original absorption spectrum, since derivative spectra resolve the overlapping band into its components and the higher order derivatives make the separation better. After the effect of line width was reduced by the derivative method, advanced computational methods such as chemometrics or multivariant analysis was employed to categorize bacteria. In Fig. 8.2, it is shown that factor analysis categorized approximately 80 bacteria into groups. For instance, when second derivative spectral data in the range of 1,425 and 1,485 cm^{-1} are taken into account, bacteria formed two groups, gram-negative organisms (C$_1$) and gram-positive organisms (C$_2$). Moreover, when a combined range 3,000 – 2,800 and 1,500 – 700 cm^{-1} is used, bacteria are separated into four clusters in the plot of the score of the factor 3 against that of factor 2. These groups are the *Legionella pneumophila* strains (C$_1$), aeromonades

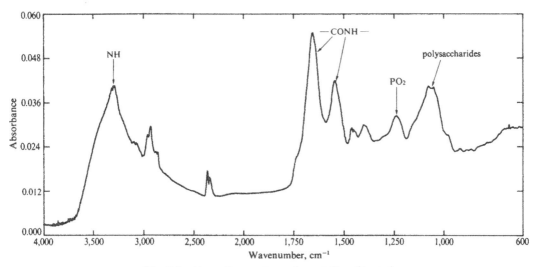

Fig. 8.1 Absorption spectrum of a cast film of bacteria.

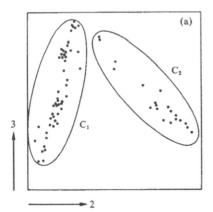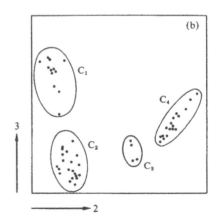

Fig. 8.2   Plot of factorial eigenimage 2 *versus* 3, showing several groups.   (a) using spectral range 1,424–1,482 cm⁻¹.
(b) using spectral ranges 3,000–2,800 and 1,500–700 cm⁻¹.

and pseudomonades (C_2), the capsular antigen carring *Escherischia coli* strains (C_3), and the *Salmonella typhimurium* variants from an epidemic cause (C_4), respectively, as shown in Fig. 8.2.

In this section, the technique to obtain such IR spectra of bacteria is explained.   Bacteria grown on a culture medium is harvested with a spatula and dispersed in a drop of clean water. The water is then transferred to a ZnSe plate and dried to form a transparent sheet of bacteria on the plate.   The chart shows an absorption spectrum (a) and first fourth (b–e) derivative spectra in the frequency range of 2,000 – 620 cm⁻¹.   Since a sharp weak spectral feature is emphasized more with higher order derivativation, one will note far larger numbers of bands in higher derivative spectra.   However, since even a weak spectral feature overlapping the spectra of a bacteriaum becomes more conspicuous in the derivative spectra, one must be very careful to remove spectral contributions from water vapor and carbon dioxide.   Although spectral features below 800 cm⁻¹ are not important in categorizing bacteria, we have included this range to show that a weak signal of carbon dioxide at 667 cm⁻¹ seen in (a) becomes a very strong feature in the derivative spectra (c), (d), and (e).

To observe IR spectra of bacteria directly from culture media using an IR microscope is also reported.[5]

1. W.W. Coblentz, *Bul. Natl. Bur. Standards*, **7**, 619 (1911).
2. R. Stair and W.W. Coblentz, *J. Res. Natl. Bur. Standards*, **15**, 295 (1935).
3. J. W. Riddle, P.W. Kabler, A. Kenner, R.H. Bordner, S.W. Rockwood and H.J.R. Stevenson, *J. Bacteriol.*, **72**, 593 (1956).
4. D. Naumann, *Mikrochim. Acta*, **1**, 373 (1988).
5. P. L. Lang and S. Sang, *FACSS abstract*, #760 (1994).

Chart 8A

<Experimental conditions>
4 cm^{-1} resolution, DTGS detector, 20 scans

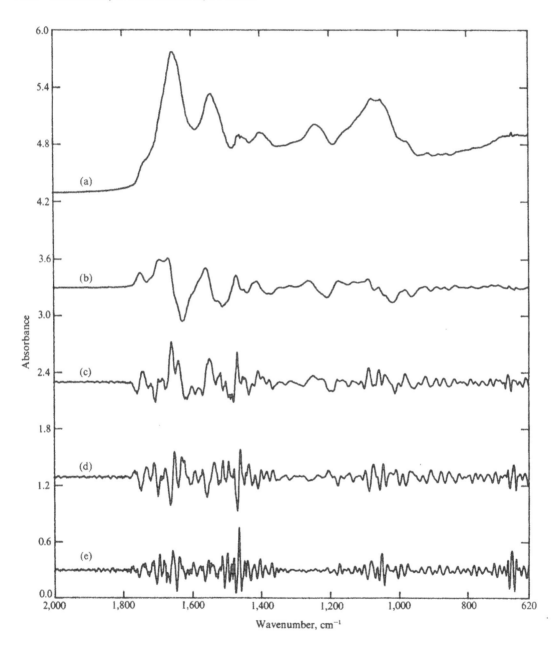

8B	Denaturation of Protein	Diffuse Reflection

IR spectroscopy is often utilized to study the higher order structure of biological samples.   In this section, spectral change accompanying the denaturation of protein is shown.   *S*-carboxymethylkeratin (SCMK) is a protein obtained from wool microfibrin after scales have been removed.   One of the present authors (E.N.) and others studied the change in the higher order structure of SCMK by thermal treatment, since there is interest in SCMK as a possible material for use in the human body.   Since a powder sample was available, the diffuse reflection technique was used for this study.   All diffuse reflection spectra were observed at room temperature for SCMK isolated from Merino sheep wool after the sample was heated for 15 minutes at several different temperatures in the sample cup of the high-vacuum high-temperature diffuse reflectance cell.   Starting from 70°C, the treatment temperature was raised to 230°C as shown in the chart.   It was observed that the frequency of the amide-II band was shifted from 1,564 cm^{-1} to 1,535 cm^{-1} by thermal treatment, especially above 120°C.

As shown in Fig. 8.3, a spiral form (α form or α-helix) (a) and an extended chain (β-form or β-pleat) (c) as well as a random coil (b) are known in the second order structure of protein.   After a number of IR studies on protein higher order structure, several peaks constituting the amide-II band are assigned to the above-mentioned specific structures, as shown in Table 8.1. However, Table 8.1 indicates that some of the SCMK bands are found in different frequencies.[1,2,3]

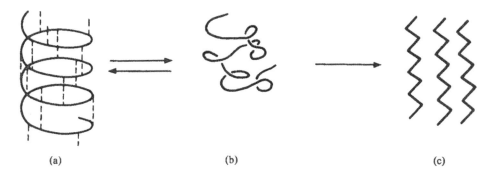

(a)   (b)   (c)

Fig. 8.3   Second order structure of protein and random coil.

TABLE 8.1   Relationship between amide-II band and higher structure

Structure	Polarity	cm^{-1}	Keratin
Random coil		1,535	
α-helix	//	1,516 (weak)	
	⊥	1,546	1,560
β-parallel	//	1,530	1,500
	⊥	1,550	
β-antiparallel	//	1,530	1,500

Chart 8B

<Experimental conditions>
2 cm^{-1} resolution, DTGS detector, 30 scans

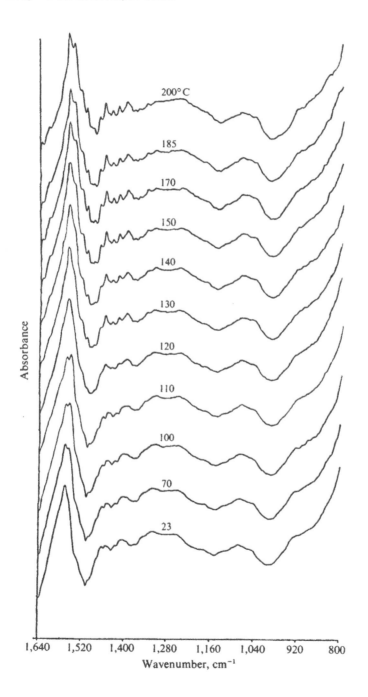

As is clear from the chart, the most intense peak is the perpendicular absorption due to an α-helix (1,560 cm^{-1}) at room temperature.   As the treatment temperature is raised, a peak at 1,535 cm^{-1} due to the random coil increases while the intensity of the α-helix band diminishes, the change being accute at a temperature of about 120°C.   It was concluded the α-helix structure changes to the random coil structure by heating.

1. T. Miyazawa and E.R. Blout, *J. Am. Chem. Soc.,* **83**, 712 (1962).
2. E.G. Bendit, *Biopolymers,* **4**, 539 (1966).
3. J. Koga, K. Kawaguchi, E. Nishio, N. Ikuta, I. Abe and T. Hirashima, *J. Appl. Polym. Sci.,* **37**, 2131 (1989).

8C	Structure Change by Drawing SCMK[1]	Transmission

Keratin, similar to myocin, epidermin, and fibrinogen, is a fibrous protein and is mostly of α-helix form.   Keratin is an elastic material which can be stretched but resumes its original length when the stretching force is removed.   However, it will yield to the force when it is overdrawn. This section describes an infrared study of the structure change due to drawing, using SCMK film as an example.   Transmission IR (unpolarized) spectra were measured for samples stretched 0, 10, 20, 30, 35, 40, 45, 50, 55, 60, 70, 80, 90, 100, 120, 160, and 200% of the original length.   The chart contains some of these spectra.   It can be seen that some of the peaks appearing in the central portions of the amide-I ($\sim 1,650$ cm^{-1}) and amide-II ($\sim 1,550$ cm^{-1}) bands decrease in intensity as the film is stretched, and both bands look like totally absorbing bands when the stretching distance exceeds 50%.   A simulation method (see Section 11.3) was used to determine the intensities of the $\alpha_{//}$, $\alpha_{\perp}$, $\beta_{//}$, $\beta_{\perp}$, and random coil contributions to the amide-II* band for all of the observed spectra (see Table 8.1 in the previous section).   Fractions of α-helix, β-sheet, and random coil were estimated from the spectral intensities of the α-helix (sum of $\alpha_{//}$ and $\alpha_{\perp}$), β-sheet (sum of $\beta_{//}$ and $\beta_{\perp}$), and random coil absorptions, all intensities of which being normalized to a unit film thickness.   In Fig. 8.4, each fraction was plotted as a function of % stretching.   The

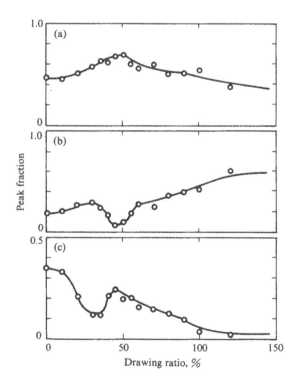

Fig. 8.4   Changes in peak fractions of α-helix (a), β-sheet (b), and random coil (c) due to drawing.

---

* See previous section for the frequencies of these bands.

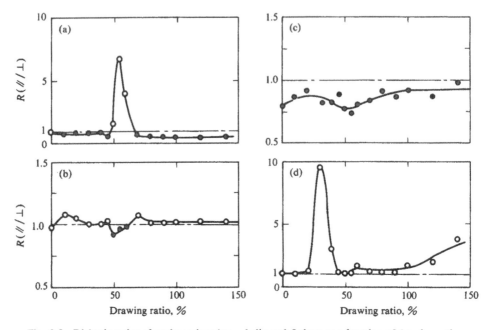

Fig. 8.5   Dichroic ratios of peaks assigned to α-helix and β-sheet as a function of drawing ratio.

stretching range from 25% to 60% shows anomalous behavior.   In this range, the fraction of the α-helix slightly decreases and the fraction of the β-sheet increases largely at the expense of the random coils.   This implies that the polymer chains are realigned to form β-sheets when the random coil polymer chains are stretched, the change is schematically depicted in Fig. 8.3.

Dichroic ratios were determined in order to investigate the anomaly seen in the 25% – 60% stretching range.   Dichroic ratios of the $\alpha_\perp$, $\alpha_{//}$, $\beta_\perp$, and $\beta_{//}$ bands were plotted *versus* % stretch in Fig. 8.5.   The dichroic ratios of $\alpha_\perp$ and $\beta_{//}$ changed abruptly in the 25% – 60% stretching range. The experimental data have been explained in terms of the transient orientations of small domains of helix and sheet during this range of drawing.   A similar phenomenon was discovered for poly(vinyl chloride) in 1961.[2]

1.   J. Koga, K. Kawaguchi, E. Nishio, K. Joko, N. Ikuta, I. Abe and T. Hirashima, *J. Appl. Poym. Sci.*, **37**, 2131 (1989).
2.   M. Tasumi and T. Shimanouchi, *Spectrochim. Acta*, **17**, 731 (1961).

Chart 8C

<Experimental conditions>
4 cm^{-1} resolution, DTGS detector, 16 scans

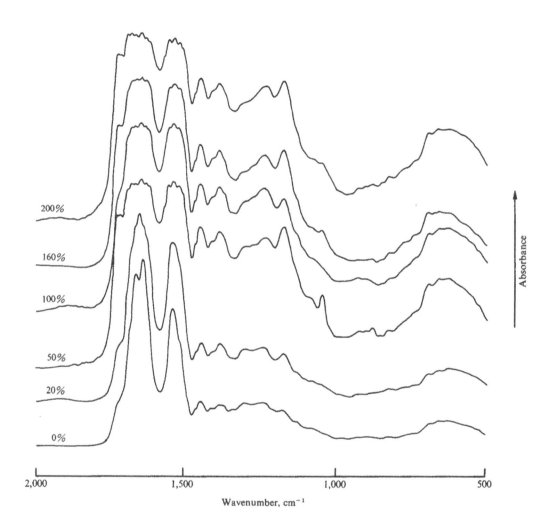

| 8D | **Pigment of Butterfly Wing**[1] | **Diffuse Reflection** |

A yellow pigment found in the wing of a butterfly native to South America was investigated. The diffuse reflection technique was chosen because diffuse reflection of the radiation was expected at the surface of the wing, since the butterfly wing does not have a uniform film structure but a complicated structure with pores and squamous or scale-like cells. Spectrum (a) of the chart was taken from the yellow part of the wing placed in a micro-sampling cup of the diffuse reflection accessory, the piece of wing being covered with about 1 mm thick fine KBr powder. Although some of the spectral features seem to coincide with the IR spectrum of the protein as shown in Section 8B, there are many sharp overlapping lines. Spectrum (b) is a difference spectrum obtained from spectrum (a) and a reference spectrum, which is obtained from the wing piece following extraction of the yellow pigment with 70% ethanol. The majority of the absorption bands compared with reference spectrum (c) of 3-hydroxykynurenine [$HONH_2$-$C_6H_3COCH_2CH(NH_2)COOH$],[2] indicates that virtually all of the bands in spectrum (b) coincide with the reference spectrum, leading to the conclusion that the yellow pigment of the butterfly wing is solely 3-hydroxykynurenine. The authors investigated the pigment of other butterfly wings and found that the brown pattern in these wings was composed of melamine.

1. E. Nishio and H. Naoki, Tokyo Symposium an Applied Spectrometry, (1986).
2. Y. Umebachi, *Zoological Science,* **2**, 163 (1985).

Chart 8D

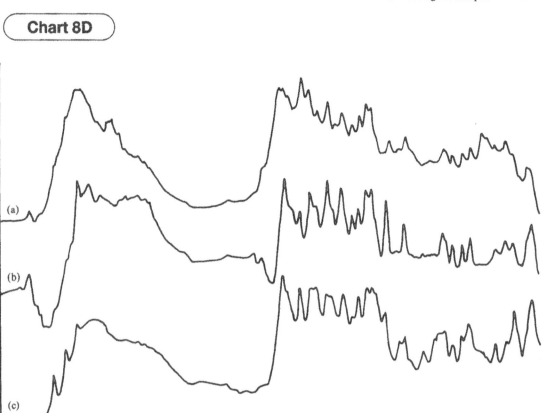

(a)

(b)

(c)

000   2,000   1,000   400

Wavenumber, cm^{-1}

9. Analysis of Semiconducting and Superconducting Materials

9.1 IR Spectra of Si and GaAs Wafers

The semiconducting material which serves as substrate for integrated circuits (IC) is mostly silicon single crystal and is supplied to IC manufacturers in the form of thin (0.3 – 0.6 mm) wafers. Normally one of the surfaces is optically polished, but both sides may be polished. Thus, in a majority of the cases, samples brought to the analytical laboratory are wafers of this simple type or those treated in the process of IC production. The treatment includes the deposition of an epitaxial layer, formation of thin silicon oxide layer, and printing of the electric circuit. The Si single crystal is manufactured by a zone-melting method to achieve ultra high purity. During the manufacturing process, trace quantities of oxygen and carbon as impurities may not be removed from the crystal. In addition, to implement designated electric properties, the Si is often doped with small amounts of elements such as boron (B) or phosphorus (P). Si doped with boron has positive holes and is called a p-type semiconductor, while Si doped with phosphorus has a free electron and is called an n-type semiconductor. The measurement of impurities is important in the IC industry to characterize the material. Undoped Si wafer shows a high electrical resistivity and transmits IR radiation. As the concentration of dopant increases, the electrical conductivity increases due to free holes or electrons. These holes and electrons allow electric conduction. As their concentration increases the transmission of IR radiation decreases.

Galium arsenside, GaAs, is also an important material in the semiconductor industry and is supplied as wafer also. In this section, the reader will find several examples of the analyses of Si and GaAs wafers.

9.1A	Absorption and Reflection Spectra of Si Wafer (Ⅰ)

In this section, the features of transmission and reflection spectra of Si wafer are described. Spectra in the chart are a transmission (a) and a reflection (b) spectrum of a low conductivity Si wafer with boron as the dopant.   The most prominent feature of these spectra is an interference fringe pattern.   However, the reader will note broad absorption bands at $1,110$ cm^{-1} due to oxygen and 600 cm^{-1} due to lattice vibration in both spectra.   Except for the fact that the peak intensities of the reflection spectrum are weak compared with those of the transmission spectrum, spectra (a) and (b) have the same pattern.   This is in strong contrast with absorption and reflection spectra of other samples shown in this volume (see Fig. 2.16, Sections 11.5A and 11.5B).

Consider the reflection and refraction at the boundry between air ($n_1 = 1$) and Si ($n_2 = 3.4$). The reflectivity at the boundary, $R = (n_1 - n_2)^2/(n_1 + n_2)^2$, is 29.98%.   Thus, as shown in Fig. 9.1, due to multiple reflections, the transmitted radiation is composed of radiation which has passsed through the wafer once and three times.   On the other hand, the reflected radiation is composed of components reflected at the surface and that which has passed through the wafer twice, the former contribution being *ca.* $2/3$ and the latter *ca.* $1/3$ of the total reflected light.   Although one-third of the reflected radiation carries the absorption component, two-thirds of the reflected radiation is the specular reflection component, which shows negligible spectral features.   The reflection spectrum of the Si wafer becomes an absorption spectrum as a result.

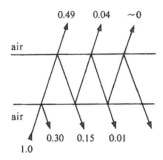

Fig. 9.1   Multiple reflections at the air–Si boundary.

Chart 9.1A

<Experimental conditions>
0.5 cm⁻¹ resolution, DTGS detector, 20 scans

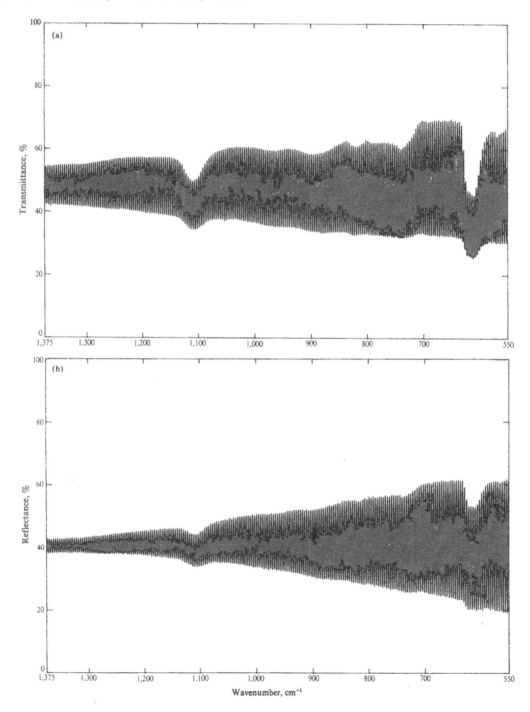

| 9.1B | Absorption and Reflection Spectra of Si Wafer (II) |

Spectra in the chart are the absorption (a) and reflection (b) spectra of a highly doped Si wafer (380 μm thick). Both spectra, especially the reflection spectrum, look noisy but the noise-like lines are an interference fringe pattern which are not well resolved in the graphics. Due to free carrier absorption, transmission of the light is small, *ca.* 6%T at 4,000 cm^{-1}, and it becomes zero below 1,200 cm^{-1}. Therefore, it is impossible to measure the absorption intensities of bands due to oxygen (1,106 cm^{-1}), or carbon (604 cm^{-1}), and other impurities in order to determine the concentration of relevant elements. However, since the reflection spectrum (b) has enough band intensity over the entire frequency range, the reflection spectrum represents the preferred choice for the determination of impurities. Unfortunately, the features at 1,106 and 604 cm^{-1} seen in the undoped Si wafer (Section 9.1A) do not appear in the reflection spectrum (b). It will be evident from Fig. 9.1 that the radiation below 1,200 cm^{-1} is absorbed and does not return to the surface. Thus, the reflection spectrum does not have any absorption component which carries information regarding impurities.

Figure 9.2 is a reflection spectrum of an epitaxial layer (3.5 μm) composed of undoped silicon formed on the same Si wafer. A high frequency interference fringe pattern, expanded in Fig. 9.3, and a low frequency fringe pattern starting at around 2,000 cm^{-1} are observed. With the assumption that the refractive index of the epitaxial layer is the same as pure Si, $n = 3.42$, the thickness of the epitaxial layer is calculated from the low frequency fringe pattern to be $d = 3.5$ μm using the following equation,

$$d = m/2n(\tilde{\nu}_1 - \tilde{\nu}_2) \text{ and } \tilde{\nu}_1 = 1,970 \text{ cm}^{-1}, \ \tilde{\nu}_2 = 660 \text{ cm}^{-1}, \text{ and } m = 3.$$

Provided that the high frequency fringe pattern is due to the Si wafer, the above equation was

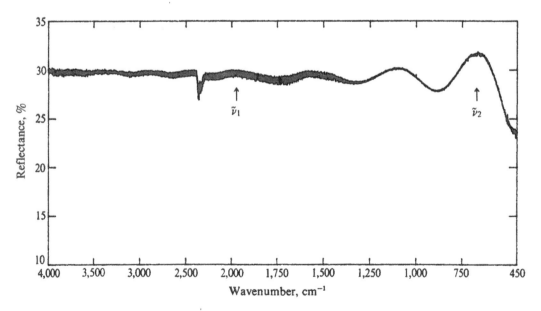

Fig. 9.2 Reflection spectrum of high conductivity Si wafer.

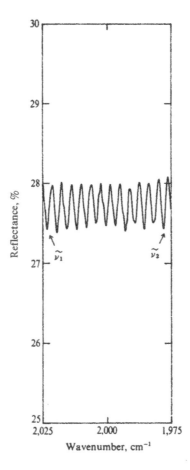

Fig. 9.3 Expansion of Fig. 9.2 highlighting the interference fringe pattern.

utilized to estimate the refractive index of the doped Si wafer. Inputting the thickness of the wafer, $380\,\mu$m, we obtained a refractive index of 3.40. Since the calculated thickness of the epitaxial layer and the refractive index of silicon agree well with those observed, the analysis of the interference fringe pattern is reasonable.

It is clear that the determination of impurities is impossible in the case of a highly doped Si wafer, although it is possible to determine the thickness of the epitaxial layer formed on such doped Si wafers.

Chart 9.1B

<Experimental conditions>
0.5 cm^{-1} resolution, DTGS detector, 20 scans

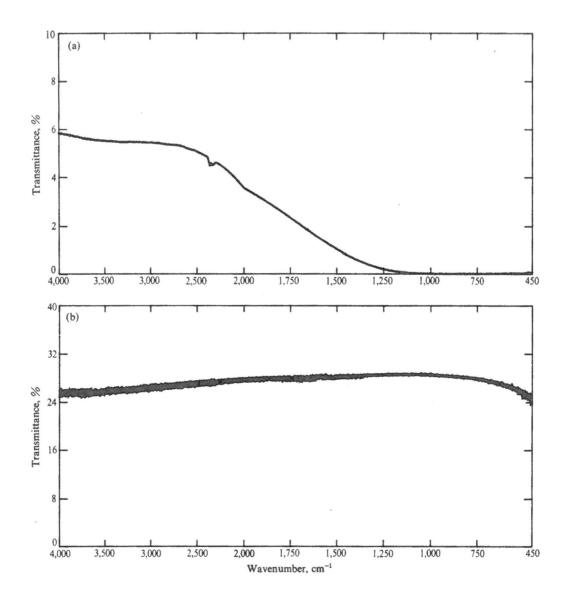

9.1C	Removal of Interference Fringe Pattern (I)

As shown by spectrum (a) of Section 9.1A, a well-developed interference fringe pattern is associated with the transmission and reflection spectra of Si wafers.   The interferogram which results in a spectrum with an interference fringe pattern shows a spike as marked by an arrow in the chart, the pattern in the annotation being the expanded portion in the vicinity of the spike.   If the interferogram is truncated just before the spike and Fourier-transformed, the interference fringe pattern will disappear as shown in Fig. 9.4.   However, an interferogram with small retardation results in a poor resolution spectrum.   Removal of the spike from the interferogram by means other than truncation will remove the interference fringe pattern without losing high resolution.

A spectrum of oxygen impurity, shown in Fig. 9.5 in absorbance units, was obtained from a modified interferogram, in which all of the energy values across the spike were changed to 0. The interference fringe pattern (see spectrum (a) of Section 9.1A), whose original spacing and amplitude are more or less uniform and for which the amplitude is about two times larger than the absorption band at $1,106 \, cm^{-1}$, disappears in Fig. 9.5.   However, unevenly spaced small ripples with an irregular amplitude were generated instead.   The appearance of these ripples results from the discontinuity in the interferogram at the retardation where the values were set to 0.   Thus, although it is possible to remove the interference fringe pattern from the spectrum, this method to remove the spike from the interferogram is not a perfect method.

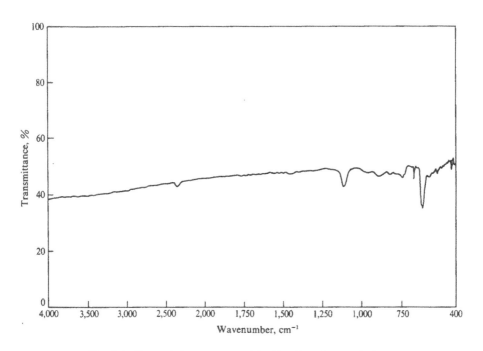

Fig. 9.4   Transmission spectrum of the Si wafer at low resolution.

Another method has been reported in which a portion of the interferogram immediately adjoining the spike is substituted for the spike region or a background interferogram section is used.   This method is effective in removing the interference fringe pattern.   However, since the replaced portion of the interferogram is not expressed in true values, the modified interferogram gave a spectrum almost identical to the spectrum shown in Fig. 9.5.

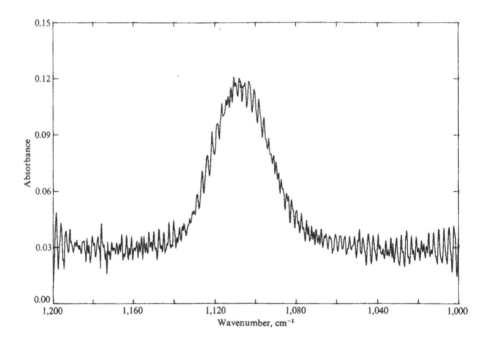

Fig. 9.5   IR spectrum of the Si wafer after removal of spike in interferogram.

Chart 9.1C

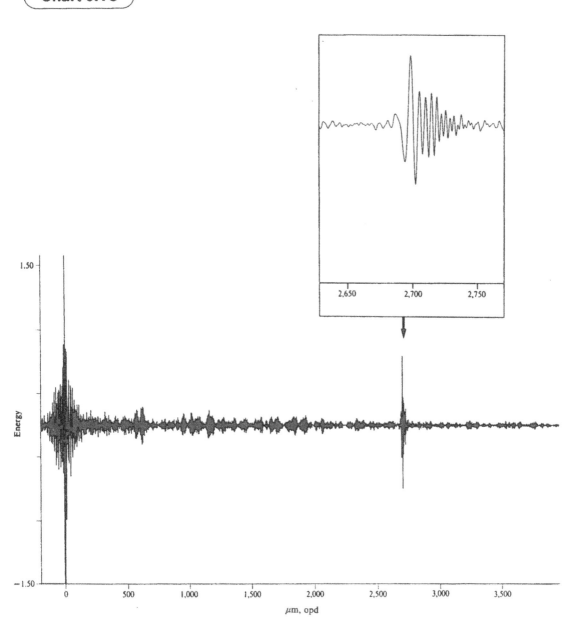

9.1D | Removal of Interference Fringe Pattern (II)

The IR transmission spectrum of a GaAs wafer, both surfaces of which were optically polished, was measured with $0.4\,cm^{-1}$ resolution. Because the thickness and the refractive index ($n = 3.26$) of the wafer are such, the spacing of the interference fringe pattern was about $0.4\,cm^{-1}$, as shown in chart (a). It is well known that due to carbon impurities relatively high resolution such as $0.5\,cm^{-1}$ and better is needed to measure the absorption band in Si and GaAs wafers for quantitative analysis. Therefore, it was necessary to find a suitable technique to remove the interference fringes obtained when the spectrum was measured at $0.4\,cm^{-1}$.

Since the spacing of the interference fringes is nearly equal to the instrument resolution, the spike in the interferogram corresponding to the fringe pattern was expected to exist near the end of the interferogram. Thus, the resolution of the FT-IR measurement was slightly changed to make the retardation slightly smaller and a series of spectra was taken.

Chart (b) was obtained when the resolution was set to $0.44\,cm^{-1}$. The interference fringe pattern had almost completely disappeared. Compared with the spectrum taken with $0.40\,cm^{-1}$ resolution, it is obvious that quantitation of the peak intensity is much more reliable in the absence of the fringe pattern. However, since the residual fringe pattern may degrade the accuracy of the peak height or peak area determination, a difference spectrum was calculated between the original spectrum and a frequency-shifted spectrum, the magnitude of the shift being an exact multiple of one full wave of the fringe pattern. Fig. 9.6 shows that the difference spectrum is further free from the residual fringe pattern.

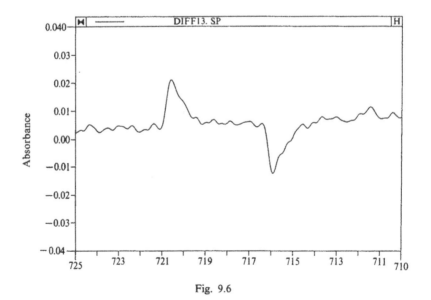

Fig. 9.6

Chart 9.1D

<Experimental conditions>
0.40 cm^{-1} resolution for spectrum (a) and 0.44 cm^{-1} resolution for spectrum (b), DTGS detector, 36 scans

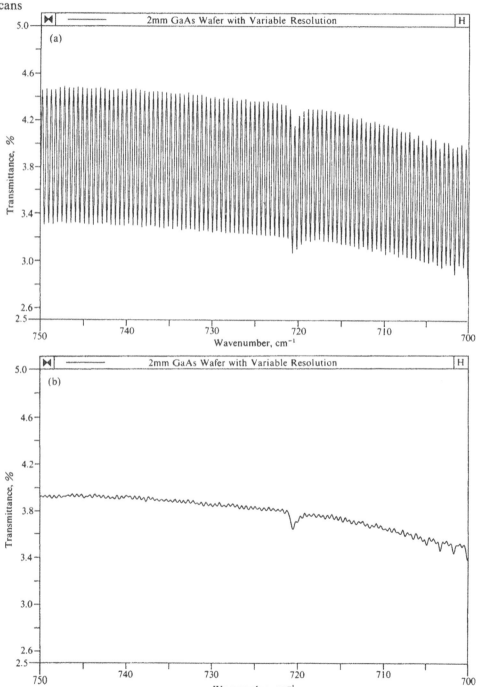

9.1E	Impurities in Si Wafer	Transmission

Figure 9.4 in Section 9.1C shows absorptions at 1,106 and 604 cm^{-1}.   The peak at 1,106 cm^{-1} provides information about oxygen interstitially incorporated into the Si crystal to form non-linear Si–O–Si bonds, and the peak at 604 cm^{-1} is assigned to the lattice vibration of Si crystal. The peak at 604 cm^{-1} is a moderately broad peak and a quite sharp peak of the carbon impurity in the crystal lattice will overlap this peak at 604 cm^{-1}.   In addition to these two impurities, several impurities are known to show absorption bands in the low frequency range below 500 cm^{-1} at low temperatures such as liquid helium temperature.   The spectrum in the chart is an example of such a spectrum measured at 11 K.   Some of the examples are: 548 cm^{-1} (Ga), 472 cm^{-1} (Al), 382 cm^{-1} (As), 320 cm^{-1} (B), 316 cm^{-1} (P), and 294 cm^{-1} (Sb).

Although oxygen gives rise to a broad single peak at 1,106 cm^{-1} at room temperature (Fig. 9.7(a)), this peak shows a high frequency shift and splits into four narrow absorption bands at low temperatures, such as 11 K, as shown in Fig. 9.7(b).   Thus, low temperature measurements are preferable for the determination of multiple impurities.

When the concentration of impurities is to be calculated,[1] one must use the peak intensity taking into account multiple reflection effects.   Transmittance (T) and absorbance (α) are given in Eqs. 9.1 and 9.2, respectively, when the multiple reflection is taken into account*

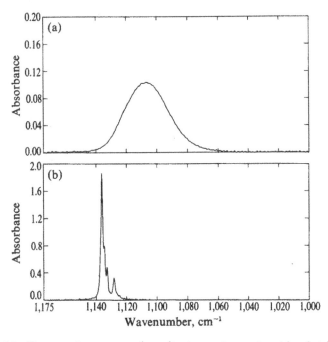

Fig. 9.7   IR spectra due to oxygen impurity at room temperature (a) and at 11K.

---

* With $e^{-\alpha d} = X$, $T = (1-R)^2 X/(1-R^2 X^2)$.   Thus, $R^2 T X^2 + (1-R)^2 X - T = 0$.

Solving the equation for X, one obtains $X = (-(1-R)^2 \pm \sqrt{[(1-R)^4 + 4T^2 R^2]})/2TR^2$.

The plus sign must be used to make the X value positive.   Finally, we obtain $\alpha = (1/d)\ln(1/X)$.

Chart 9.1E

<Experimental conditions>
4 cm^{-1} resolution, DTGS detector, 36 scans

<Accessory>
Air Product He Cryostat to cool the sample to 11 K.

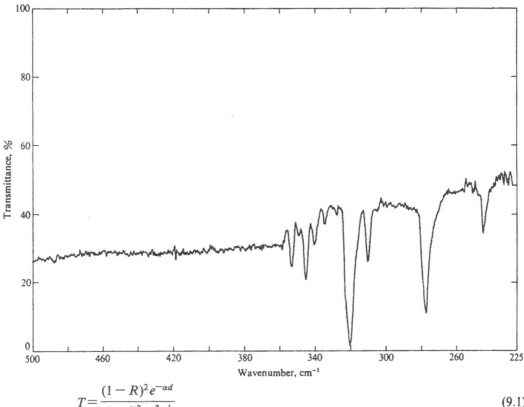

$$T = \frac{(1-R)^2 e^{-\alpha d}}{1 - R^2 e^{-2\alpha d}} \tag{9.1}$$

$$\alpha = \frac{1}{d} \ln \frac{2TR^2}{\{(1-R)^4 + 4R^2 T^2\}^{1/2} - (1-R)^2} \tag{9.2}$$

where $R = 0.2997^*$ and d is the sample thickness.   However, in the IC industry, another value, 0.2995, is used for the reflectance, R.   After the correction for the effect of lattice vibration (intensity $\alpha_e = 0.85$ cm^{-1}), peak intensity in the difference spectrum with a reference material with the same thickness is used.

In the case of highly reflecting materials such as Si, Ge, and GaAS, a method calculating absorbance simply as $A = -\log(1/T)$ should not be used.

1.   (a) W.Kaiser, P.H. Keck, and C.F. Lange, *Phys.Rev.*, **101**, 1264 (1965); (b) B. Pajot, *Analyst*, **5**, 32 (1977); (c) H.J. Hrostowski, R.H. Kaise, *Phys. Rev.*, **107**, 966 (1957); (d) ASTM F 1,188 – 88

---

* Reflectance at the boundary between air and silicon ($n = 3.42$) is $R = (1 - 3.42)^2/(1 + 3.42)^2 = 0.2997$.

9.1F	SiO₂ Layer on Surface of Si Wafer (I)	ATR

We attempted to obtain the IR spectrum of a SiO_2 layer formed on the surface of a Si wafer[1]. A 5,000 Å thick SiO_2 layer formed on a 3.5-μm thick phosphorus-doped silicon epitaxial layer on a nitrogen-doped high conductivity Si wafer. Since this nitrogen-doped Si-wafer did not transmit IR radiation below 1,200 cm^{-1} where the SiO_2 spectrum is expected, an ATR method was utilized. Sectrum (a) in the chart was measured with a Ge IRE and an incident angle of 45°. Three peaks at 1,138, 1,065, and 801 cm^{-1} with intensities and shapes corresponding of those of SiO_2 were observed indicating that the correct experiment was performed. On the other hand, spectrum (b) taken with KRS-5 IRE and 45° incident angle was significantly different from spectrum (a). Since the penetration depth in the case of the KRS-5/SiO_2 combination (*ca.* 2 μm at 1,000 cm^{-1} as given by $d_p = 0.2\lambda$; see Table 2.1 in Chapter 2) is larger than that of Ge/SiO_2 ($d_p = 0.6\,\mu$m for 1,000 cm^{-1} radiation), a stronger absorption is expected for the KRS-5 experiment. However, the Restrahlen-band-like feature at 1,300 cm^{-1} cannot be explained in terms of the large penetration depth only. As shown above the penetration depth of the 1,200 – 800 cm^{-1} radiation is far larger than the thickness of the SiO_2 layer in the case of the KRS-5 observation. Therefore, the ATR condition of the present KRS-5/SiO_2/Si system should rather be judged by a case like KRS-5 ($n = 2.37$)/Si($n = 3.42$), in which the condition of total reflection is never achieved. In other words, the spectrum (a) is not an ATR spectrum, but is closer to a surface reflection measurement. On the other hand, Ge/SiO_2/Si seems to barely fulfill the total reflection criterium, when the thickness of the SiO_2 is comparable to the penetration depth, although the optical nature of the substrate must be taken into account for a more precise description of the spectrum. Thus, one must select the correct experimental conditions when the SiO_2 layer on the Si wafer is measured by an ATR method, because the thickness of the SiO_2 layer is sometimes smaller than the penetration depth.

1. (a) K.H. Beckmann and N.J. Harrick, *J. Electrochem. Soc.,* **118**, 614 (1971); (b) A. Hartstein, D.J. DiMaria, D.W. Dong and J.A. Kucga, *J. Appl. Phys.,* **51**, 3860 (1980); (c) A. Hartstein and D.R. Young, *Appl. Phys. Lett.,* **38**, 631 (1981); (d) I.W. Boyd and J.I.B. Wilson, *J. Appl. Phys.,* **53**, 4166 (1982); (e) J.E. Olsen and F. Shimura, *Appl. Phys. Lett.,* **53**, 1934 (1988).

Chart 9.1F

<Experimental conditions>
4 cm^{-1} resolution, DTGS detector, 64 scans

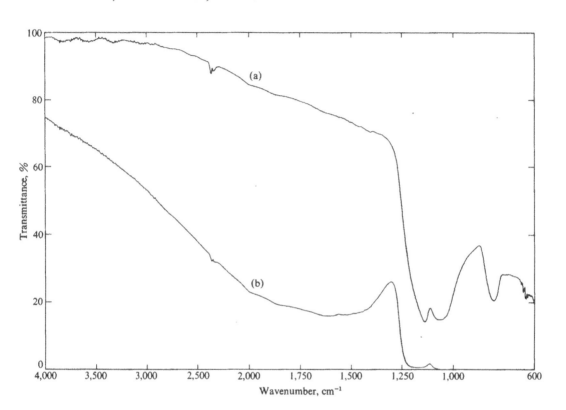

| 9.1G | SiO$_2$ Layer on Surface of Si Wafer (II) | ATR |

Shimura[1] reported that not only the optical properties of the IRE and SiO$_2$ but those of the Si substrate should also be included if a full understanding of the ATR spectrum of SiO$_2$ formed on Si wafers is to be reached.   According to Harrick,[2] the effective thickness of the sample observed with ATR (refractive index of IRE $=n_1$ and incident angle of θ) technique for parallel and perpendicular polarizations is given in Eqs. 9.3 and 9.4, respectively.   A thin sample with a refractive index of n_2 whose thickness satisfies the criterion, $2\pi d/\lambda<0.1$, is assumed to be supported on a thick substrate (refractive index $=n_3$) as follows:

$$d_{e//} = \frac{4n_{21}\cos\theta[(1+n_{32}^4)\sin^2\theta - n_{31}^2]}{(1-n_{31}^2)[(1+n_{31}^2)\sin^2\theta - n_{31}^2]}\,d \tag{9.3}$$

$$d_{e\perp} = \frac{4n_{21}\cos\theta}{(1-n_{31}^2)}\,d \tag{9.4}$$

where d and λ are thickness of the sample and wavelength of the IR radiation, respectively. Relative refractive index is given by $n_{ij}=n_i/n_j$ ($i=1, 2, 3$ and $j=1, 2$).

Since a total reflection condition must be achieved even for the ultimate case of $d=0$ (reflection at the IRE/Si boundary where the refractive index of Si is larger than that of SiO$_2$), Ge IRE ($n_1=4.00$) and an incident angle of 60° were selected to study SiO$_2$ thin layer on Si wafer.[*1]   Use of a multiple internal reflection accessory, in which a 2 mm thick 52 mm long IRE with a 60° angle of incidence is used, provided 15 reflections.   Thus, Eqs. 9.3 and 9.4 yield the effective thickness of SiO$_2$ ($n_2=1.46$), $d_{e//}=1{,}560d$ and $d_{e\perp}=39d$, which indicates that the ATR method has a large amplification effect in this case.

The spectrum shown in the chart is an ATR spectrum of a naturally ozidized Si-wafer,[*2,*3] using a Ge IRE and an incident angle of 60° as described above.   Some of the reported absorption bands related to oxygen in the Si crystal are deposited silica at 1,225 cm^{-1}, interstitial oxygen atoms at 1,107 cm^{-1}, so-called "thick oxide" and so-called "thin oxide" at 1,080 and 1,050 cm^{-1} as well as cristobalite (semi-stable crystal of SiO$_2$)[3] at 1,230 cm^{-1}.   In this chart, although cristobalite was not observed, interstitial oxygen and a thin oxide were.

It was found that the ATR method with Ge IRE at a 60° angle of incidence is the most advantageous method to study thin oxide layers on Si wafers.

1.   J.E. Olsen and F. Shimura, *Appl. Phys. Lett.*, **53**, 1934 (1988).
2.   N.J. Harrick and F.K. duPre, *Appl. Opt.*, **5**, 1739 (1966).
3.   (a) F. Shimura, H. Tsuya and T. Kawamura, *Appl. Phys. Lett.*, **37**, 483 (1980); (b) G. Haas and C. D. Salzberg, *J. Opt. Soc. Amer.*, **44**, 181 (1954); (c) A. Hjortsberg and C. G. Grandvist, *Appl. Opt.*, **19**, 1694 (1980).

[*1] A critical angle is 58.5° for Ge($n=4.00$) and Si($n=3.42$) boundary
[*2] Sample courtesy of Kyushu Denshi Kinzoku Co. Ltd.
[*3] I.W. Boyd and J.I.B. Wilson (*Appl. Phys.*, **53**, 4166 (1982)) state that the thickness of oxide layer is *ca.* 22 Å two weeks after production.

Chart 9.1G

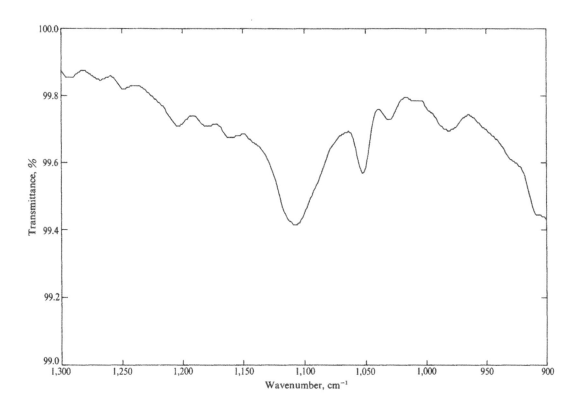

9.2   IR Spectrum of a Superconductive Material

9.2A	Microspectroscopy of a Super-Conductive Material	Microscope

The relationship between superconduction and crystal structure is discussed here based on the IR study of an organic superconducting material.   The sample is a single crystal of di[bis-(ethylenedithiolo)tetra-thiafulvalene]–bis(isothiocyanato)cuprate (I), $(BEDT-TTF)_2[Cu(CNS)_2]$, which shows superconductivity below 10.4 K.[1]   Since the purpose of the study is to understand superconductivity in terms of crystal structure, the observation of an IR spectrum must be carried out with the single crystal and not with a KBr pellet.   Since the crystal size is *ca.* $1 \times 2 \times 2$ mm, a normal transmission method is unusable, leaving the choice of reflection spectroscopy of the crystal using an IR microscope as a possibility.

The spectra in Chart (a) are specular reflection spectra of the sample with polarized radiation parallel to the b and c axes of the crystal as indicated in the chart.   Although a frequency range of 27,000 – 450 cm^{-1} was studied, only the 6,000 – 450 cm^{-1} range was studied with the FT-IR microscope system and the rest of the range was studied with a UV/VIS microspectrometer.[2]   Chart (b) displays the conductivity calculated using Kramers-Kronig analysis of reflection spectra.   A sharp intense peak at 880 cm^{-1} was assigned to a Raman active C–S symmetric vibration which became IR active in the crystal due to electro-vibronic interaction.   Peaks at 1,150 and 1,320 cm^{-1} are assigned to CH bending vibrations, while the 2,050 and 2,100 cm^{-1} bands are assigned to the C≡N stretching vibrations of the –NCS group.   As shown in both parallel and perpendicularly polarized reflection spectra in chart (a), a large dispersion due to free electrons is seen in the reflection minimum at around at 5,000 cm^{-1}.   The free electron dispersion model describes this dispersion phenomenon as

$$\varepsilon(\omega) = \varepsilon_{core} - \omega_p^2/\omega(\omega + i/\tau)$$

where ω_p is a plasma frequency and ε is a dielectric constant. τ is the relaxation time of carriers and $i = \sqrt{-1}$.   As explained in Chapter 2, the complex refractive index of the sample is obtained through Kramers-Kronig analysis of the reflection spectra.   Since the complex dielectric constant, $\hat{\varepsilon}$, is related to the complex refractive index, $\hat{n}$, as $\hat{\varepsilon} = \hat{n}^2$, ε_{core}, ω_p, and τ, are obtained by Kramers-Kronig analysis of the reflection spectrum.   Sugano *et al.* estimated the band gap for the superconduction as 0.73 eV and 0.40 eV along the b and c axes, respectively.[4]   The crystal structure of this material is shown in Fig. 9.8.   Those interested in IR spectra of electroconducting materials are urged to study solid state optics.[5]

1.   H. Urayama, H. Yamochi, G. Saito, K. Nozawa, T. Sugano, M. Kinoshita, S. Sato, K. Oshima, A. Kawamoto and J. Tanaka, *Chem. Lett.,* **1988**, 55.
2.   T. Sugano, K. Yamada, G. Saito and M. Kinoshita, *Solid State Commun.,* **55**, 137 (1986).
3.   For instance, Charles Kittel, "*Introduction to Solid State Physics,* Chapter 8, Wiley, New York (1986).
4.   T. Sugano, H. Hayashi, H. Takenouchi, K. Nishikida, H. Urayama, H. Yamochi, G. Saito and M. Kinoshita, *Phys. Rev. B,* **37**, 9100 (1988).
5.   Charles Kittel, *Introduction to Solid State Physics* (Third Edition), Wiley, New York (1966).

Chart 9.2A

<Experimental conditions>
8 cm^{-1} resolution, DTGS detector combined with IR-PLAN microscope, 100 scans

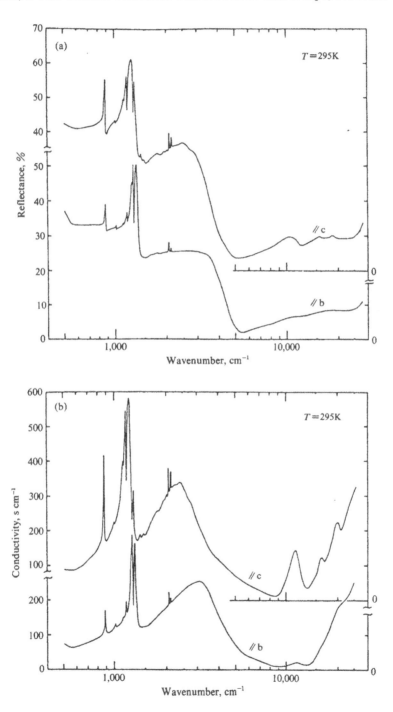

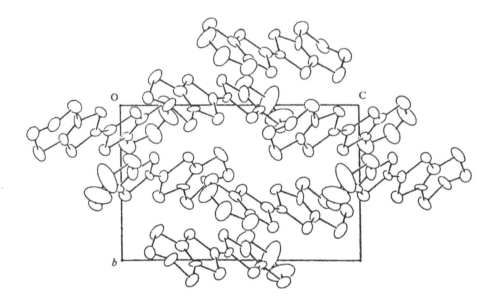

Fig. 9.8   Crystal structure of (BEDT–TTF)₂[Cu(CNS)₂].

10. Hyphenated Techniques

The term "hyphenated technique" signifies a combined technique in which one analytical method is interfaced to another. Thus, when the FT-IR system is chosen as one of the techniques, the counterpart can be gas-chromatography (GC), liquid chromatography (LC), thermal gravimetric analysis (TG), or dynamic mechanical analysis (DMA). Among the following four hyphenated techniques, GC-FTIR, LC-FTIR, TG-FTIR, and DMA-FTIR, the first three are rather commonly used. Some examples of hyphenated techniques are presented in this chapter. The major advantage of hyphenated techniques is the significant increase in the amount and quality of information about the sample.

10.1 LC-FTIR Techniques

LC-FTIR technique suffers from a technical problem, *i.e.*, strong solvent absorptions. In order to provide the full frequency range of observation, it is necessary to use a flow cell with a small path length. However, the concentration of the solute is normally quite low and the LC peak has a narrow width when typical HPLC conditions are employed. Thus, it is difficult to get a good signal-to-noise, full range spectrum of a solute, although it is possible to detect some of strongly absorbing peaks for limited qualitative and quantitative purposes. The situation becomes worse in the case of reversed phase LC analysis, especially when buffer salts are added since these show strong absorptions, a fact which further increases the frequency range of total blackout. In addition, it is practically impossible for the FT-IR system to compensate for the solvent absorptions when a gradient technique empolying mixed solvent systems is employed.

In spite of these limitations, the LC-FTIR technique must be used in some cases where the GC-FTIR technique is not compatible with the large molecular weight of the analytes. Gel Permeation Chromatography (GPC) is one of the operational modes of LC. GPC utilizes a column packed with porous material so that the molecules are trapped inside the pores. Since smaller size molecules are easily trapped inside the pores and large molecules cannot penetrate pores, the largest size molecules come out of the column first. Thus, the GPC method separates molecules by molecular size, or molecular weight difference under experimental conditions. Since the polymer has a certain molecular weight dispersion giving rise to a rather broad peak in the chromatogram, the chromatographic resolution does not need to be high as ordinary LC conditions to identify low molecular weight solutes. It is possible to inject a large portion of sample, using even a single solvent, since high resolution is not necessary. Taking advantage of these aspects of GPC operational conditions, one may expect some of the difficulties mentioned above to be eliminated or reduced. In addition, since only a few IR absorption peaks are needed to characterize polymers in the case of single polymer analysis (regardless of whether the sample is a homopolymer or copolymer), it is not neccessary to obtain the entire IR frequency range.

GPC-FTIR technique benefits the study of polymer mixtures by aiding the identification of components and additives and characterization of polymers as deduced from the microstructure of the sample as a function of molecular weight. Examples shown in this section are GPC-FTIR analyses which can be performed without requiring any specialized techniques except a heated cell and transfer tubing in some cases.

| 10.1A | GPC-FTIR Analysis of Poly(ethylene) Additives |

One may notice a few weak peaks, which are not assigned to poly(ethylene) (PE), in the IR spectra taken from commercially available PE products such as shopping bags, packing materials, and pellets.   Peaks at *ca.* 3,700 and 1,740 cm^{-1} are examples of these peaks. Although the peak at 1,740 cm^{-1} can be attributed to some kind of oxidized PE produced during production, it may also be due to one of the additives in PE.   Since all of these extra peaks are weak in intensity, it is not easy to investigate the chemical structures of these additives.   In this section, the analysis of the PE additives by means of Gel Permeation Chromatography-FTIR (GPC-FTIR) technique is described.

Additives were extracted for 12 hours from a 3 g portion of commercially available PE pellets using a Soxlet extractor with hexane as solvent.   The extracted solution was injected into an HPLC system equipped with TSK Gel G 2000 H8 (7.5 mm × 60 cm) column.   The flow rate of the solvent (CCl$_4$) was 1.0 ml/min.   The effluent of the GPC system was transferred to a flow cell (1.0 mm path length) and returned to a refractive index (RI) detector of the GPC system.

Figure 10.1 shows the chromatogram constructed from interferograms showing total absorption intensity at each time slice.   The IR chromatogram shows that there are four components in the extracts, the first one of which shows an IR spectrum quite similar to that of PE.   Indeed, the first component is low molecular weight PE, which is so-called "poly(ethylene) wax" added to PE to improve its physical properties.   Spectra (a), (b), and (c) observed from the remaining three peaks (2, 3, and 4) are shown in the chart.   Spectrum (a) is a spectrum typical of acrylic esters (C=O stretching at 1,734, –C–O– ester bands at 1,250 and 1,160 cm^{-1}) and it was assigned to an acrylic polymer which is added to suppress water molecules from traversing the polyethylene sheet.   Spectrum (b) shows bands due to aromatic compound (a weak peak at 3,100 cm^{-1}, peaks at 1,500 and 1,450 cm^{-1} due to the aromatic ring) with a sharp strong

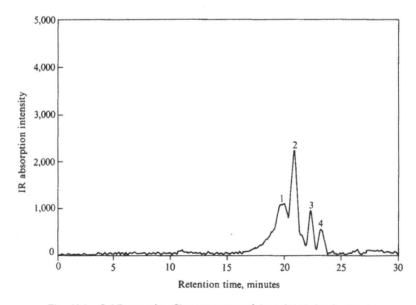

Fig. 10.1   Gel Permeation Chromatogram of the poly(ethylene) extract.

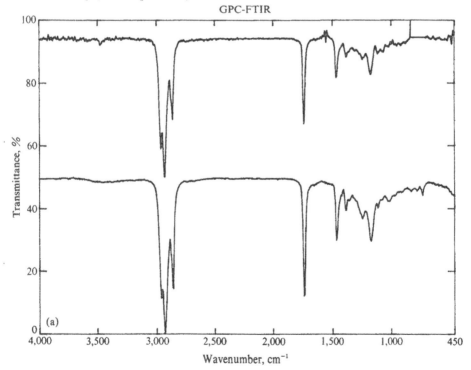

Fig. 10.2   Molecular structure of tetrakis[methylene-
              3(3,5-di-*tert*-butyl(4-hydroxyphenel)pro-
              pionate]methane.

Fig. 10.3   Molecular structure of 2,6-di-
              tert-butyl-*p*-cresol (BHT).

absorption at 3,650 cm^{-1}.   This suggests that the compound is similar to 2,6-di-*tert*-butylphenol, which is commonly used as an antioxidant.   In addition, the compound has a carbonyl group, most probably an ester structure, as peaks at *ca.* 1,750 and 1,163 cm^{-1} reveal.   The best match was found with tetrakis[methylene-3(3,5-di-*tert*-butyl(4-hydroxyphenyl)propionate]methane (Fig. 10.2) from the Sadtler IR Library.   Spectrum (c) also showed typical 2,6-di-*tert*-butyl phenol features, and based on the best match to the Sadtler IR Library, we assigned the IR spectrum to BHT (2,6-di-*tert*-butyl-*p*-cresol (Fig. 10.3), which is the fundamental antioxidant used for the stabilization of PE.

Chart 10.1A

<Experimental conditions>
8 cm^{-1} resolution, DTGS detector, automatic data acquisition with GC-IR software
Observed spectra, (a), (b) and (c) (upper spectrum in each chart) together with the best match to the Sadtler IR Library (lower spectrum).

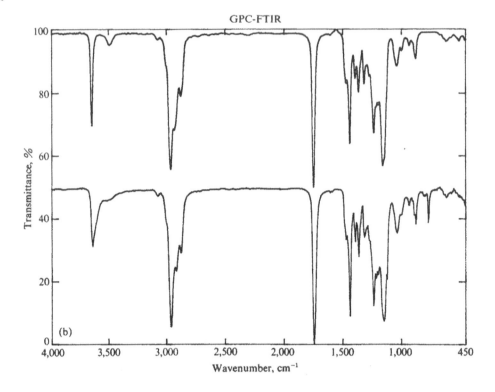

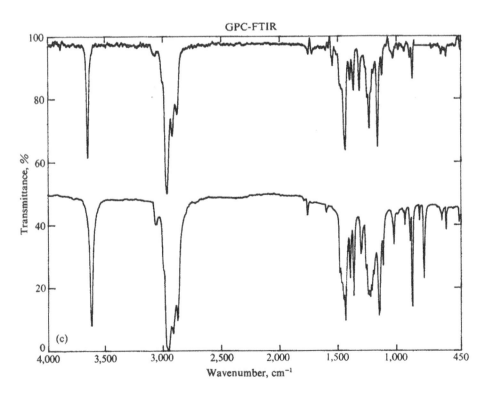

| 10.1B | Characterization of Poly(propylene) by GPC-FTIR |

Gel Permeation Chromatography is very important in quality control in the polymer industry, because it is a convenient method to provide molecular weight and molecular weight dispersion with ease and speed.   Since many of the physical and chemical properties depend not only on molecular weight and molecular weight dispersion but also on the microstructure of polymer molecules, it is beneficial to study the chemical structure of polymers as a function of molecular weight.

In this section, the characterization of poly(propylene) (PP) with high temperature GPC-FTIR technique is shown.   PP is soluble only in certain solvents, such as 1,2,4-trichlorobenzene (TCB) or o-dichlorobenzene (ODCB), at elevated temperatures, e.g., 130°C.   The eluent from a commercially available high temperature GPC system was transferred to a temperature-controlled flow cell (quartz window with 1 mm path length) and returned to the RI detector built into the GPC system via a temperature-controlled transfer tube.   The peak area of methyl group vibration at 2,960 cm^{-1} relative to that of methylene group vibration at 2,920 cm^{-1} was analyzed to determine the ethylene contents of ethylene-propylene copolymer as a function of molecular weight.   Fig. 10.4 shows the IR spectra of the CH stretching vibrations for three copolymers with different ethylene content as well as ethylene and propylene and propylene homopolymers. Since the peak due to a methyl group at 2,960 cm^{-1} disappears when the ethylene content exceeds 50%, a deconvolution technique (see Chapter 11) was used to determine the intensities of the methyl and methylene group absorptions.   By measuring more than 20 predetermined standard polymers, we obtained a linear calibration to cover the ethylene content from 0 to 100%.

Figures (a), (b), and (c) in the chart show the ethylene content of an ethylene-propylene random copolymer, a block copolymer, and a mixture of polyethylene and polypropylene as a function of

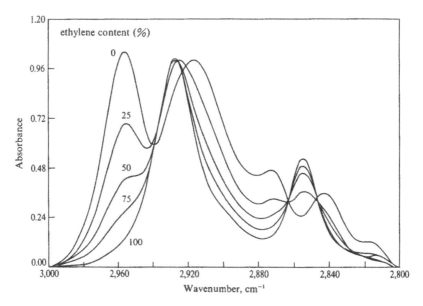

Fig. 10.4   CH stretching vibrations of ethylene-propylene copolymers with different compositions including homo-
       polymers.

molecular weight.   The ethylene content is larger in the low molecular weight fraction of the random copolymer.   This is consistent with the production method and the nature of the polymerization reaction.   Since the propagation center of polyethylene has high reactivity and less stability, the ethylene-propagating species is deactivated earlier than the propagation center of polypropylene, yielding higher ethylene content at lower molecular weight.   The block copolymer is manufactured in such a way that the propylene is polymerized first, followed by the addition of ethylene to the reaction vessel, so that the ethylene is polymerized on an active end of polypropylene chain.   Thus, the high molecular weight fraction has higher ethylene content.   In the case of the mixed copolymer, the difference between the molecular weight dispersions of poly(ethylene) and poly(propylene) determines the pattern and thus no unique pattern is produced; a typical pattern is shown in Fig. (c) of the chart.

It is clear from the chart that these three different kinds of copolymers are easily distinguished.

Chart 10.1B

<Experimental conditions>

8 cm^{-1} resolution, medium band MCT detector, spectra accumulated for 30 sec was stored as time-sliced data (GPC conditions) temperature: $130°C$, solvent: TCB, injection: 1 ml portion of 0.3% TCB solution of polymer, GPC solumn: TSK Gel G 2000 H8 (7.5 mmϕ × 60 cm)

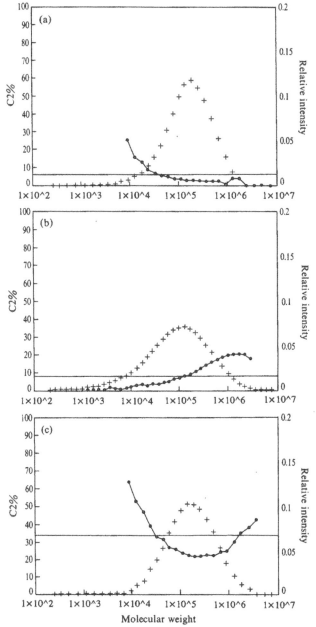

·－·－·－·: ethylene (C₂) contents (%), +++++: GPC curve as determined from infrared absorption intensity,
———: average C₂ contents.

10.1C | Characterization of Poly(ethylene)[1] by GPC-FTIR

Although the physical properties of poly(ethylene) depend naturally on its molecular weight, molecular weight dispersion, density, and crystallinity, it is known that the environmental stress crack resistance of so-called linear low density poly(ethylene) (LLDPE) is improved when the degree of short chain branching is increased[2] in the higher molecular weight fraction.   A similar effect due to short chain branching is expected for high density poly(ethylene) (HDPE).   Since a methyl group is the terminal group of the short chain branchings of LLDPE and HDPE, the content of methyl groups yields the number of short chain branchings after a correction has been made for a methyl group at one terminus of the PE molecular chain.   The number of short chain branchings was determined as a function of molecular weight for several LLDPE and HDPE samples together with several low density PE (LDPE) samples for reference.   The instrumental system described in the preceding section can be used to characterize poly(ethylene) without any modification, because the system determines the ratio of methyl groups *versus* methylene groups.

Figures in the chart show the degree of short chain branching as a function of molecular weight (dots connected by lines) and the GPC curve as determined from the CH stretching vibration peak area (3,000 – 2,800 cm^{-1}) for (a) a LLDPE polymerized in the laboratory using a Ziegler-Natta catalyst, (b) a commercially available LLDPE, and (c) a commercially available HDPE used for thin wrapping film.   Sample (a) shows a monotonic decrease in the degree of short chain branching as the molecular weight increases, while sample (b) gives almost a constant degree of short chain branching over the entire molecular weight range.   According to the literature and a patent granted to some PE manufacturers, sample (b) must have better crack resistance than sample (a).   According to Hosoda,[3] LLDPE polymerized with a soluble catalyst such as ethyl

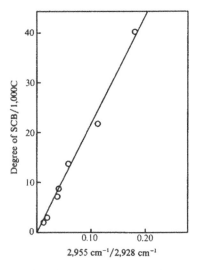

Fig. 10.5   Relationship between numbers of total methyl groups per 1,000 carbon atoms and peak ratio, A2965/A2920, of deconvoluted IR spectrum of 1,2,4-trichlorobenzene solution of polyethylene.

1. K. Nishikida, T. Housaki, M. Morimoto and T. Kinoshita, *J. Chromatography*, **517**, 209 (1990). and references therein.
2. Asahi Kasei Co. Ltd. Japan Patent SHO61–26049 and SHO60–26050 (1985).
3. S. Hosoda, *Polym. J.*, **20**, 383 (1988).

Chart 10.1C

<Experimental conditions>

8 cm^{-1} resolution, medium band MCT detector, accumulation of 30-second scans as a time-slice file, GC-IR software for automatic data accumulation (GPC conditions) temperature of GPC system: 130°C, column: TSK Gel G2000 H8, injection volume: 1 ml aliquot of 0.3% TCB solution of PE, flow rate: 1 ml/min, solvent: TCB

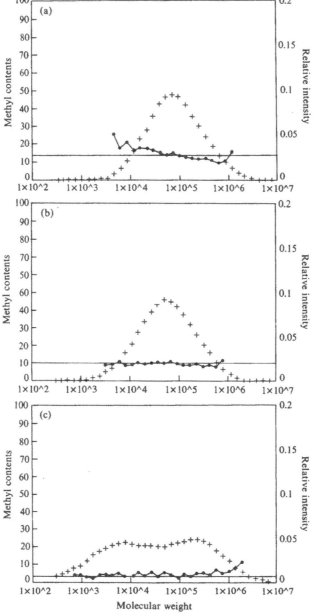

·—·—·: methyl contents/1000 carbon, +++++: GPC curve as determined from infrared absorption intensity,
————: average methyl contents/1000 carbon.

aluminum sesquichloride/vanadium chloride has virtually no variation in short chain branchings as a function of molecular weight.

Sample (c), which is the most successful HDPE for thin wrapping film, shows a very characteristic double peak GPC curve and an almost constant number of short chain branches or perhaps a slight increase with increase in molecular weight.

In this experiment, since the methyl group absorption at 2,960 cm^{-1} is too weak to appear in the PE spectra, a deconvolution technique was used to reduce the linewidth, so that the weak feature due to the methyl group could be separated from the shoulder of the intense peak due to methylene groups.   As shown in Fig. 10.5, the system was calibrated to determine the total number of methyl groups per 1,000 carbon atoms in the range from 2 to 40.

10.2   GC-FTIR Technique

The GC-FTIR technique, in which the FT-IR system is used as a detector for gas chromatography, has inferior sensitivity compared with GC-mass spectrometry (GC-MS). However, the GC-IR technique has a definite advantage over GC-MS, since gas state IR spectra offer more discrete information on the chemical structures, particularly of isomers than mass spectra.   Since the GC-IR analysis is more or less straightforward depending on the nature of the sample, we are concerned with rather special applications utilizing the Curie-point pyrolyzer-GC-FTIR and Super Fluid Extraction-GC-FTIR methods in this section.   Standard GC-FTIR techniques are described elsewhere.[1]

1.   Robert White, *Chromatography/Fourier Transform Infrared Spectroscopy and Its Application*, Marcel Dekker, New York, (   ).

<table>
<tr><td>10.2A</td><td>Pyrolysis GC-FTIR Analysis of Silane Coupling Reagent of Glass Fiber</td></tr>
</table>

Silane coupling reagent at the boundary between glass fiber and plastic plays an important role in the characteristics of glass fiber-reinforced plastics (GFRP). Therefore, it is worth investigating the stoichio-chemical features of GFRP in addition to the mechanical characteristics for evaluation of GFRP as an industrial material. Analytical methods for the determination of silane coupling reagents bound on the surface of glass fibers were throughly investigated by Ishida.[1] They used a diffuse reflection technique to analyze the reagents from different sample morphologies such as woven glass mat and silica-gel powder. One of the present authors (E.N.) and others[2] employed a transmission method (Section 4A) and a KBr pellet technique (Section 4B) for a woven glass mat, and showed that it is possible to observe quantitatively reliable IR spectra of the reagent except in the 1,300 – 800 cm^{-1} region where the strong SiO$_2$ Restrahlen band prevents an accurate measurement.

The presence of silica prevents high repeatability IR measurements, so removal of the silica from the sample is desirable. Pyrolysis GC-FTIR technique was therefore employed to remove the silane coupling reagent from the non-volatile silica substrate, even though thermal decomposition of the reagent is inevitable.

In order to find a suitable pyrolysis temperature, thermal gravitational analysis (TGA) was performed on silica fibers treated with a silane coupling reagent, γ-anilinopropyl trimethoxylsilane. Fig. 10.6 shows the weight loss curve *versus* temperature from which the pyrolysis temperature was chosen to be 750°C.

The figure shown in the left of the chart is a chromatogram detected with the GC detector. An IR spectrum (d) observed from the peak at 4 min of retention time is assgned to C$_6$H$_5$–NH–CH$_2$–CH$_2$–CH$_3$, which corresponds to a fragment of the silane coupling reagent. The IR absorption

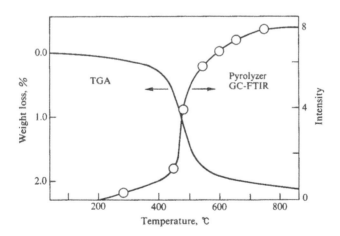

Fig. 10.6   Weight loss curve of glass fiber treated with silane coupling reagent.

1. (a) H. Ishida and J.L. Koenig, *J. Colloid Interface Sic.*, **64**, 565 (1978); (b) S. Naviroj, S.R. Cullar, J.L. Koenig and H. Ishida, *ibid.*, **97**, 308 (1984); (c) S.R. Cullar, H. Ishida, J.L. Koenig, *ibid.*, **106**, 334 (1984).
2. (a) E. Nishio, N. Ikuta, T. Hirashima and J. Koga, *Appl. Spectrosc.*, **43**, 1159 (1989); (b) N. Ikuta, Z. Maekawa, H. Hamada, S. Yoshioka, E. Nishio and T. Hirashima, to be published.

Chart 10.2A

<Experimental conditions>
8 cm^{-1} resolution, narrow band MCT detector, automatic accumulation based on GC-IR
software

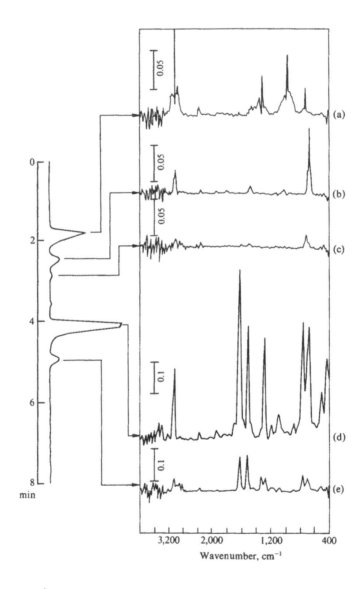

intensity integrated over the entire GC peak was found to be proportional to the concentration
of the reagent in the reaction mixture used to treat the glass fiber.   The integrated IR intensity
is also given in Fig. 10.6 *versus* temperature, showing good coincidence with the results of the
TGA.

10.2B	Supercritical Fluid Extraction GC-FTIR Analysis of Basil*

Supercritical Fluid Extraction (SFE) process employs a supercritical fluid to extract solid materials, *i.e.* a substance, for example, Xenon or carbon dioxide, held above its critical pressure and temperature. Supercritical fluids show some combined properties of gas and liquid. Supercritical fluid has both diffusivity to penetrate various matrixces like gas and dissolution properties like liquid.[1]

In this section, SFE-GC-IR application to basil, a common spice, is shown. One gram of basil was extracted with SFE using supercritical carbon dioxide (250 atms at 60°C). Effluent was transferred to an injection device heated to 250°C in a GC-FTIR system. After the carbon dioxide was decompressed and strong absorptions at 2,360 and 667 cm^{-1} disappeared, the background spectrum was observed for GC-IR measurement.

Chart (a) shows a Gram-Schmidt chromatogram of basil by SFE-GC-IR method. Spectrum in chart (b) presents one of the minor species observed in the present work. A previous work[2]

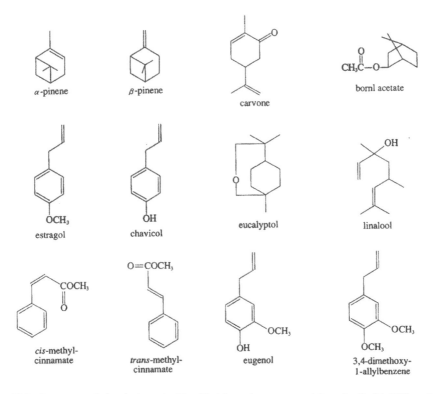

Fig. 10.7   Structures of chemical species identified from those extracted from basil with SFE method.

1. *Analytical Supercritical Fluid Chromatography and Extraction,* (M.L. Lee and K.E. Markedes, Eds.), Chromatography Conferences, Inc., Provo, Utah, 1990.
2. S.B. Hawthorne, D.J. Miller, M.S. Krieger, *J. Chromatogr. Sci.,* **27**, 347 (1989).

* Data supplied by Dr. Brian J. McGrattan and Dr. Gregory L. McClure of Perkin-Elmer Corporation.

using SFE-GC-Mass-spectrometry finding of 1,8-cineole, $C_{10}H_{18}O$, estragole, euginol and a series of sesquiterpenes with the molecular formula $C_{15}H_{24}$.   All of these materials were confirmed in this study.   In addition, the $C_{10}H_{18}O$ peak was identified as linalool.   Because of the high sensitivity achieved with the present SFE-GC-IR technique, several other compounds including α-terpine, β-terpine, chvicol, carvone, bornyl acetate, *cis-* and *trans-*methyl cinnamate, and methyl euginol (see Fig. 10.7) were identified.

The total time from loading basil seeds to obtaining the automated printout of the spectral library search was 90 minutes.   This fast analysis was possible due to the fact that supercritical extraction is far faster than Soxlet extraction.   Supercritical fluid extraction was also found to cause less thermal decomposition of the extracts.

Chart 10.2B

<Experimental conditions>
Perkin-Elmer PrepMaster supercritical fluid extractor connected to Perkin-Elmer system 2000
GC-IR system
GC column: 50 meter 0.32 mm i.d. 5 micrometer film methyl silicone column

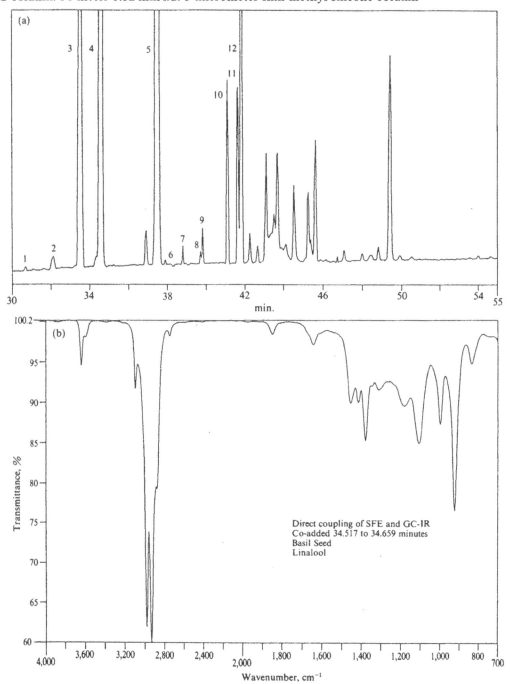

Direct coupling of SFE and GC-IR
Co-added 34.517 to 34.659 minutes
Basil Seed
Linalool

10.3   TG-IR Technique

Thermal Gravimetric Analysis (TGA) system is an analytical instrument which measures the weight loss of the sample with a micro-balance while the sample is heated in a furnace.   In the TG-IR technique, the outgas from the TGA system is analyzed on an FT-IR system.   As shown in Fig. 10.8, the sample is set in a pan hanging on a micro-balance and the pan is placed inside the furnace.   The outgas materials are transferred to a heated gas cell through a heated transfer line. This differs from the GC-IR technique, since a large volume (*ca.* 250 ml/min compared with 1 ml/ min with the GC-IR method) of purge gas is used and a large sample (max. 5 g for metals or 200 mg for organic compounds) can be used.   Therefore, the gas cell is far larger than the light pipe used for the GC-IR technique.   The gas cell used here is 8 mm diameter×6 cm in dimension.   Therefore, a DTGS detector has sufficient sensitivity to allow its use for TG-IR analysis.   Software designed for GC-IR analysis is used for data collection and data handling.

In the TG-IR experiments, the FT-IR spectrometer observes IR spectra of the outgas and records the sum of the absorptivity of the outgas materials over the entire frequency region, total absorptivity, as a function of heating temperature.   With the aid of a vector calculation software program called "Gram-Schmidt" calculation, it is possible to record the absorptivity of some special frequencies or frequency regions as a function of heating temperature as well.   The Thermogravimetric Analyzer (TGA) records the sample weight loss as a function of heating temperature.   The curve showing this weight loss *versus* temperature is called a "thermogram." Compared with the thermogram obtained from the TGA, a curve showing total absorptivity or those of special frequencies or frequency regions is called an "IR thermogram."   Although the

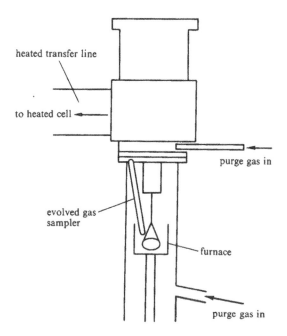

Fig. 10.8   Schematic picture of TG-IR interface. (Courtesy of Perkin-Elmer Corporation)

thermogram and IR thermogram have different ordinate dimensions, both techniques are looking at the common phenomenon of thermal decomposition of samples at the same time. Thus, the TG-IR technique provides both the weight lost and the chemical composition of the outgases due to the thermal decomposition of the samples.

It should be noted that one can stop the reaction at any stage to examine the product by elemental analysis and/or other spectroscopic methods, since the heating is controlled.

The TG-IR technique has, among others, applications in:

(1) study of the mechnisms of thermal (decomposition) reactions
(2) assignments of additives in materials
(3) detection of toxic gases generated by the thermal decomposition of materials.

In this section application to the study of thermal decomposition is discussed.

10.3A | Thermal Decomposition Study of Aspartame[1]

Aspartame is a well known artificial sweetener.   The manufacturer prohibits its use at elevated temperatures.   Since this implies that the compound is labile at high temperature, the TG-IR technique was utilized to investigate the thermal decomposition mechanism by identifying the outgas materials of the sample.   Fig. 10.9 shows (a) weight loss with time (temperature), (b) first derivative of (a), and (c) total absorptivity with time (temperature) calculated from interferograms during the TG-IR run.   TGA data, *i.e.*, weight loss data, (a) and (b) show that a two-step decomposition reaction occurs, the first step at 208°C and the second at 371°C.   At the same time, the IR data (c) show that the decomposition products generated at both steps absorb IR radiation.

In the chart, an IR spectrum accumulated for the entire temperature scan is shown in (a).   As marked in the spectrum the outgases are composed of methanol, ammonia, and carbon dioxide. Spectrum (b) was observed at the first decomposition step.   It was found that the major component of the outgas at the first step was methanol, as spectral comparison with a standard spectrum (c) in Sadtler's gas phase library reveals.

The sample was removed from the TGA pan after the first step of the decomposition was finished and an elemental analysis was performed.   Elemental analysis results support the concept that the decomposition reaction is a stoichiochemical reaction as written by

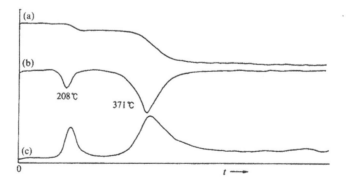

Fig. 10.9   Weight loss of Aspertame sample *versus* heating time (min) detected with (a) TGA, and (b) first derivative of (a), and (c) calculated from the total absorptivity of outgas.

*   Data provided by Dr. G. MeClure of the Perkin-Elmer Corporation.

Chart 10.3A

<Experimental conditions>
8 cm^{-1} resolution, DTGS detector, automatic data acquisition by GC-IR software

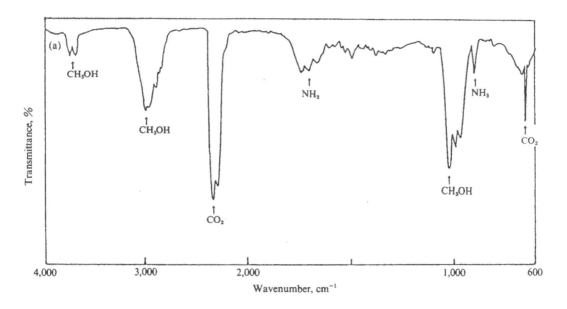

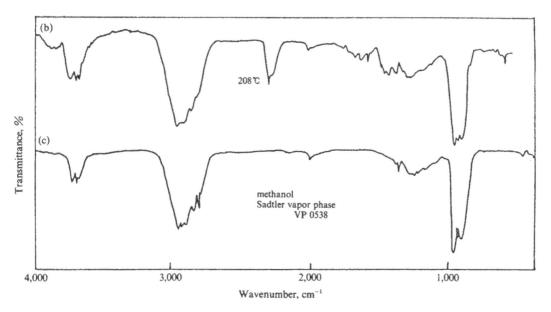

10.3B | Thermal Decomposition of Poly(vinyl chloride)

Thermal decomposition of a poly(vinyl chloride) (PVC) product was studied by means of the thermogravimetric-infrared (TG-IR) method to monitor the generation of any toxic gases. The poly(vinyl chloride) product studied in this section contained a mixed polymer (87% of PVC and 13% of ethylene-vinylacetate copolymer) and other additives such as di-octyl phthalate.

Chart (a) shows how the total absorptivity (1), the absorptivity at $1,825 - 1,750$ cm^{-1} (the carbonyl group) (2), and the absorptivity at $2,400 - 2,200$ cm^{-1} (the region for carbon dioxide) (3) of the outgases change with heating. Chart (b) shows the IR spectrum of the outgas seen during the first decomposition at 20 min (306°C) (4) together with a standard gas phase spectrum of acetic acid (CH_3COOH) (5) taken from the Sadtler gas phase library.

A number of sharp lines are observed in the frequency range of $3,100 - 2,700$ cm^{-1} in spectrum (4). These are assigned to the rotational fine structure associated with the stretching vibration of gaseous HCl. Since HCl is a diatomic molecule, this absorption band is the only normal mode of vibration. The rest of the bands in spectrum (4) showed a perfect match with the standard spectrum of acetic acid (5). Thus, HCl and acetic acid are the components which resulted from thermal decomposition at 306°C. Since the IR thermogram (2) indicates that generation of carbonyl compounds happens only at the first decomposition step, acetic acid is generated only in this decomposition step. A decomposition releasing carbon dioxide starts at *ca.* 35 min. ($T = 450$°C) as the absorption due to carbon-dioxide in the outgase starts at this temperature, as seen in the IR thermogram (3). The introduction of oxygen to the sample at 60 min ($T = 450$°C) accelerates the decomposition as the sudden increase in the absorption intensity of carbon dioxide indicates.

It was observed that HCl gas is emitted from the sample at a fairly low temperature, near 300°C, and acetic acid is also emitted at this temperature when the vinyl compound contains vinylacetate.

Chart 10.3B

<Experimental conditions>
8 cm^{-1} resolution, DTGS detector, Data acquisition automated by GC-IR software

(a)

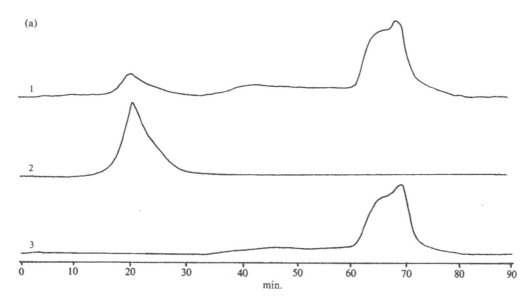

(b)

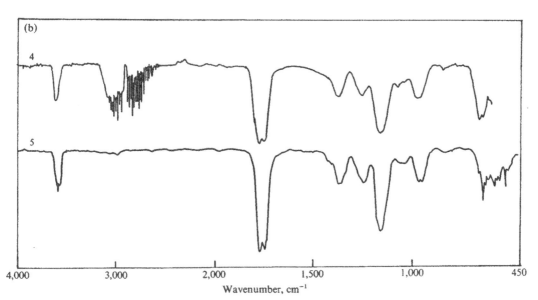

11. Application of Data Handling Software

11.1 Deconvolution

Let us start with the historical development which led to the so-called "deconvolution" technique. Any wavelike pattern observed on a measurement device is distorted due to instrument characteristics (instrument function). This process is depicted in Fig. 11.1. In addition, noise contributes to the distortion.

Thus, when a true spectrum $s(Y)$ is measured on an instrument whose instrument function is $i(Y)$ and the noise is given as $n(Y)$, the observed spectrum $o(Y)$ will be given by

$$o(Y) = i(Y) \otimes s(Y) + n(Y)$$

where $\otimes$ is convolution of the real function and the instrument function. Obtaining a real spectrum via an inverse process marked by dashed lines in Fig. 11.1 has been discussed. This process is referred to as deconvolution.[1] In the case of dispersion type spectrometers, the slit function is an essential part of instrument function when the rest of the component characteristics such as detector and optical components are removed from the observed pattern. Jones *et al.*[2] showed a method using spectral subtraction to remove the influence of the slit function on observed IR spectra. Although the object was to remove an additional line distortion from the instrument, a natural extenstion of logic led to the use of the method to remove the linewidth from the spectrum itself. Spectral resolution is apparently improved for qualitative purposes to separate overlapping components contained in a broad spectral band, regardless of the theoretical intergrity. In this section, we show two methods of deconvolution, Fourier-transform deconvolution[3] and an absorbance difference method.[4] However, we are obliged to quote a criticism[5] regarding the use of the deconvolution technique in spectroscopy as follows: As the principles of those methods will be explained in the following two sections, both methods *a priori* assume a common linewidth for all of the lines. This assumption is far from acceptance for the use of deconvoluted spectra for quantitative analysis. However, as will be shown in the following section, the deconvolution technique can be utilized for quantitative purposes if

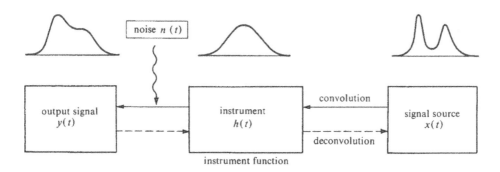

Fig. 11.1   Block diagram showing convolution (solid arrows) of instrument function and noises on a real spectral shape and deconvolution (dashed arrows) scheme.

247

sufficient numbers of standards are carefully evaluated and the use of the deconvoluted spectra gives a high correlation between peak intensity or area and concentration.

Deconvolution, properly applied, can yield additional information regarding the sample. Unfortunately, it is extremely easy to misuse deconvolution technique, and very often only experience will temper the result and its interpretation.

1. S. Minami, *Wave pattern treatments for scientific measurements,* Chap. 7, (in Japanese), CQ Shuppan, Tokyo (1986).
2. R.N. Jones, R. Venkataraghavan and J.W. Hopkins, *Spectrochim. Acta,* **23A**, 925 (1967).
3. (a) J.K. Kauppinen, D.J. Moffat, H.H. Mantsch and D.G. Cameron, *Appl. Spectrosc.,* **35**, 271 (1981); *idem., Anal. Chem.,* **53**, 1454 (1981); (b) J.K. Kauppinen, D.J. Moffat, D.G. Cameron and H.H. Mantsch, *Appl. Opt.,* **20**, 1866 (1981).
4. R.W. Hannah, Pittsbugh Conference Abstracts, Abst. No. 682 (1983).
5. B. Chase, private communication.

11.1A	Smoothing and Difference Spectroscopy Method (Hannah Method[1])

This method does not require a Fourier-transform calculation for deconvolution. The procedure is as follows.

1) Store an observed spectrum (in absorbance units).
2) Obtain its smoothed spectrum separately, using a Savitzky-Golay algorithm.
3) Subtract the smoothed spectrum from the original spectrum.
4) Adjust the subtraction factor, d, so that no part of the deconvoluted spectrum gives negative absorbance in step 3).
5) Factors for line-smoothing and spectral subtractions are determined by trial and error.
6) To conserve the area, multiply the resulting spectrum by $1/(1-d)$.

In this method, the smoothed spectrum with fewer high frequency components is subtracted from the original spectrum. As a result the difference spectrum is left with the high frequency component, that is, narrower linewidths, than the original spectrum, giving higher resolution. Spectrum (a) shows the transmission spectrum of a 1,2,4-trichloro-benzene solution of high density poly(ethylene), which has *ca.* 16 methyl groups per 1,000 carbon atoms. Since the peak intensity of the methyl group is very small, it is almost impossible to locate the asymmetric methyl CH stretching vibration. Therefore, it is impossible to quantitize the methyl group intensity *versus* CH_2 group absorptions at 2,928 and 2,858 cm^{-1}. Spectrum (b) in the chart, however, clearly demonstrates the usefulness of the deconvolution technique to isolate the weak CH_3 stretching signal from the dominant CH_2 spectral feature. Using a series of poly(ethylene) standards, all of which are calibrated with C-13 nuclear magnetic resonance (C-13 NMR), the determination of methyl group *versus* methylene group was found to be possible[2] for a range of from 2 to 40 methyl groups/1,000 carbon atoms.

1. R.W. Hannah, Pittsburgh Conference Abstracts, Abst. No. 682 (1983).
2. K. Nishikida, T. Housaki, M. Morimito and T. Kinoshita, *J. Chromatography,* **517**, 209 (1990).

Chart 11.1A

<Experimental conditions>
8 cm^{-1} resolution, DTGS detector, 30 scans, heated flow cell (1 mm path length) at 135°C

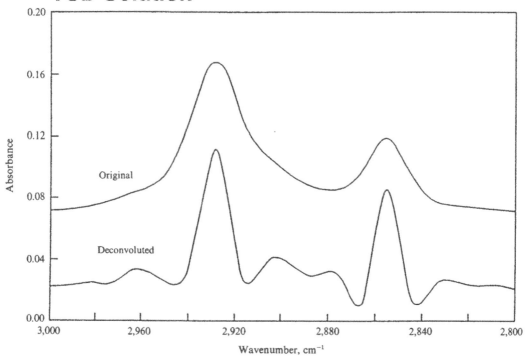

IR Spectra of Polyethylene
TCB Solution

11.1B | Fourier Deconvolution Technique

The principles of Fourier deconvolution are explained here. As discussed in Chapter 1, the interferogram of a monochromatic source without any linewidth is a constant amplitude sine wave (see Fig. 1.3(a)). On the other hand, the interferogram amplitude will decay with retardation when the radiation has a finite linewidth (Fig. 1.3(c)). It may therefore be concluded that the line shape and linewidth determine the envelope function of the decay. The interferogram shown as Fig. 1.3(c) will become Fig. 1.3(a) when the decaying interferogram is divided by its own invelope function, generating a spectrum without linewidth. Thus, if we divide the interferogram of an observed spectrum with the envelope function, which is obtained through Fourier transform of the line shape function, it is possible to modify the decay of the interferogram. The interferogram with a slow decay will yield a spectrum with a reduced linewidth. This procedure is mathematically weitten as

$$S(x) = \frac{D_s(x)}{F^{-1}(S_0(\omega))} \cdot O(x) \tag{11.1}$$

where $S(x)$ and $O(x)$ are a deconvoluted and an observed interferogram and $S(x)$ in turn provides a deconvoluted spectrum. $D_s(x)$ is an apodization function, while $F^{-1}(S_0(\omega))$ is a Fourier transform (F^{-1}) of the line shape function $S_0(\omega)$.[1] Since the Fourier transform of a Lorentzian function ($S_0(\omega) = (\sigma/\pi)/(\sigma^2 + \omega^2)$), which is one of the hypothetical line shape functions for IR spectra, is an exponential function ($F^{-1}((\sigma/\pi)/(\sigma^2 + \omega^2)) = \exp(-2\pi\sigma|x|)$), the deconvolution process involves dividing the observed interferogram by $\exp(-2\pi\sigma|x|)$ or multiplication of an observed interferogram with an exponential function ($\exp(2\pi\sigma|x|)$). Thus, although the line-shape function of the observed IR spectra would be different from Lorentzian function, most Fourier deconvolution programs are written under the assumption that the line shape of an IR absorption line may be expressed in terms of the Lorentzian function. This is justified because the purpose of the deconvolution technique is not to predict the real spectral shape but to reduce the apparent linewidth.

It was reported that the linwidth can be reduced at best to $1/2 - 1/3$ of the original linewidth by the deconvolution method. As Eq. 11.1 indicates, the deconvolution process is influenced by the apodization used. As explained in Chapter 1, apodization function affects signal-to-noise and linewidth. This is the case for the deconvoluted spectrum, as well.

Spectra (a) and (b) in the chart are the observed and deconvoluted absorption spectra of a poly(vinyl chloride) film, respectively. Many broad absorption bands are further split into several components. Although the assignment of bands in the $3,000 - 800$ cm^{-1} region is already established,[2] it is interesting to examine the discussion of the C–Cl stretching vibration bands appearing in the $700 - 600$ cm^{-1} region. In 1968, using a pattern simulation technique, Pohl and Hummel[3] reported that the triplet-like absorption band in the $700 - 600$ cm^{-1} region as seen in spectrum (a) is composed of eight absorption bands, all of which may be assigned to a different conformation of the PVC chain as shown in Table 11.1.

As Hannah's early experiment[4] and Compton's examination[5] showed, spectrum (b) confirmed six of the eight components proposed by Pohl and Hummel.

It should be noted that observation under a higher resolution condition for the FT-IR system will never resolve the unresolved structures of the observed spectrum (a) to a spectrum like spectrum (b). While the deconvolution procedure reduced the inherent linewidth of the spectrum, better instrument resolution does not remove the inherent linewidth.

Chart 11.1B

<Experimental conditions>
4 cm^{-1} resolution, DTGS detector, 16 scans

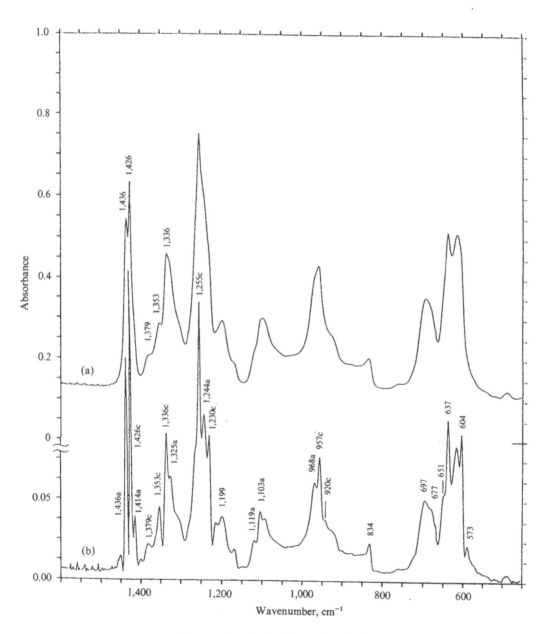

(a) amorphour band, (b) crystalline band

TABLE 11.1   Peak positions and conformation of PVC (assignment by Pohl *et al.*)

Conformation		Pohl	present
TTTT	long chain syndiotactic	603	604
TTTT	short chain syndiotactic	613	614
TGTTG'T	isotactic	624	
	isotactic	633	
TTTT	long chain syndiotactic	639	637
	syndiotactic	647	651
TGGT	syndiotactic	677	677
TGTG	isotactic	695	697

1. (a) J.K. Kauppinen, D.J. Moffat, H.H. Mantsch, D.G. Cameron, *Appl. Spectrosc.*, **35**, 271 (1981); *idem., Anal. Chem.*, **53**, 1454 (1981); (b) J.K. Kauppinen, D.J. Moffat, D.G. Cameron and H.H. Mantsch, *Appl. Opt.*, **20**, 1866 (1981).
2. (a) S. Krimm and C.Y. Liang, *J. Polym. Sci.*, **22**, 95 (1956); (b) T. Shimanouchi and M. Tasumi, *Spectrochim. Acta*, **17**, 731 (1961); (c) H.V. Pohl, J. Dexanas and D.O. Hummel, *Makromol. Chem.*, **115**, 125 (1968); *ibid*, **115**, 141 (1968)
3. H.U. Pohl and D.O. Hummel, *Makromol. Chem.*, **113**, 190 (1968); *ibid.*, **113**, 203 (1968).
4. R.W. Hannah, Pittsburgh Conference Abstracts, Abst. No 692 (1983).
5. D.A. Compton and W.F. Maddams, *Appl. Spectrosc.*, **40**, 239 (1986).

11.2   Derivative Spectra

Derivative spectra are obtained through extended use of the Savitzky-Golay algorism for digital filtering.   The Savitzky-Golay line-smoothing method takes a set of consecutive data as shown in Fig. 11.2 (a set of nine in this case).   Using the least-mean-square method, the ordinate values are fitted to 2nd or 3rd order polynomials and the ordinate of the center point is calculated as the black dot in Fig. 11.2.   Then a data region is shifted one point from the center and the above-mentioned procedure is used to calculate the ordinate value of the new center point.   This procedure is repeated from one end of the spectrum to the other.

Although the Savitzky-Golay method is a famous line-smoothing technique utilizing adopted polynomials, it is converted to calculate derivative spectra and peak positions by incorporating suitable adopted functions.

Although both methods can be used to calculate derivative spectra from any IR spectrum, regardless of the original spectra being obtained with dispersive or FT spectrometers, a method to obtain derivative spectra from interferograms using Fourier transform is known.   The Fourier transform of a single-sided interferogram which is multiplied with a function such as ax^n will give rise to an nth derivative spectrum.   For instance, a use of the parabolic function, ax^2, as an apodization function will give a second derivative spectrum.   Griffiths stated that one should avoid the use of a spectrum with unnecessarily high resolution in order to avoid noise contribution from the higher retardation when one uses the Fourier domain derivative method. He also stated that a spectrum with too low resolution will introduce side lobes in the derivative form, since the interferogram multiplied by ax^n may not decay at the position of truncation.

Another way of obtaining the derivative spectrum utilizes difference spectroscopy. In this method the difference spectrum between the original spectrum shifted by an appropriate frequency toward a higher frequency and that shifted by the same frequency toward a lower frequency becomes the first derivative spectrum.

In Section 11.2, we present two examples of the application of derivative spectra.

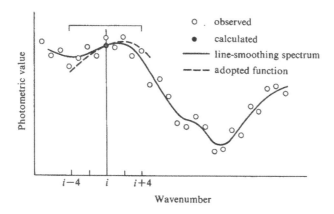

Fig. 11.2   Savitzky-Golay method for line smoothing.

11.2A   Fine Structure Analysis by Derivative Spectroscopy

The spectra in the chart are an absorption spectrum (in absorbance units) (a) and the first derivative (b) and second derivative (c) spectra of di-*sec*-butyl phthalate. The greatest disadvantage of the first dertivative spectrum exists in the difficulty of determining peak position. Peak position is a frequency where the spectrum crosses the baseline as shown in Fig. 11.3 in the case of the first derivative spectra. However, when the original spectrum shows a partially resolved band pattern such as a triplet pattern in the frequency range from 3,000 to 2,700 cm^{-1} of spectrum (a), some of the peak positions are different from the point where the first derivative spectrum crosses the baseline, for some peaks the baseline is not crossed at all. Thus, some experience is needed to utilize first derivative spectra. In this regard, second derivative spectra are easier to use to determine the peak positions, since the peak tops of the second derivative spectra coincide with the peak tops of the original spectrum.

Derivative spectroscopy has an advantage in emphasizing weak spectral features which are buried beneath dominant spectral features. At the same time, it has a disadvantage in estimating peak intensity. In derivative spectra, wide peaks yield relatively low amplitude compared with a sharp band. Note that weak peaks due to water vapor in 3,900 – 3,500 cm^{-1} become more prominent in the first derivative spectrum than in the original and the trend becomes more obvious in the second derivative spectrum. On the other hand, the slope below 550 cm^{-1} seen in the original spectrum completely disappears in the second derivative spectrum. The first derivative of a constant slope is offset but it is flat. The first derivative of a first derivative, *i.e.*,

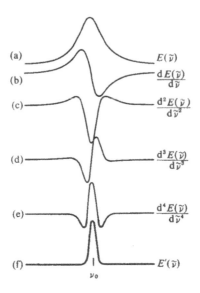

Fig. 11.3   Absorption band (a) and derivative spectra, first (b), second (c), third (d), and fourth (e) derivative. Deconvoluted band is given in (f).

the second derivative of the original function, where the region is flat has zero value.   Thus, the slope disappears in the second derivative spectrum.

Figure 11.4 is the second derivative of the original spectrum when the derivative calculation is performed on the transmission spectrum using the same procedure as for the absorbance spectrum.   Although the splitting of the carbonyl group band at 1,740 cm^{-1} into two bands is clearly observed in Fig. 11.4, the splitting is not evident in spectrum (c).   Although the derivative in the transmission unit showed extra information compared to the derivative in absorbance units, Fig. 11.4 should not be utilized for quantitative purposes, since the $-\log(d^n T / d^n \tilde{\nu})$ is not linear with concentration.

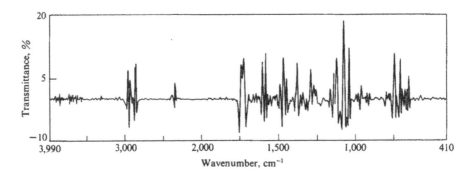

Fig. 11.4   Second derivative spectrum of di-*sec*-butyl *ortho*-phthalate.   (Calculations performed in %T units.)

Chart 11.2A

<Experimental conditions>
4 cm^{-1} resolution, DTGS detector, 16 scans

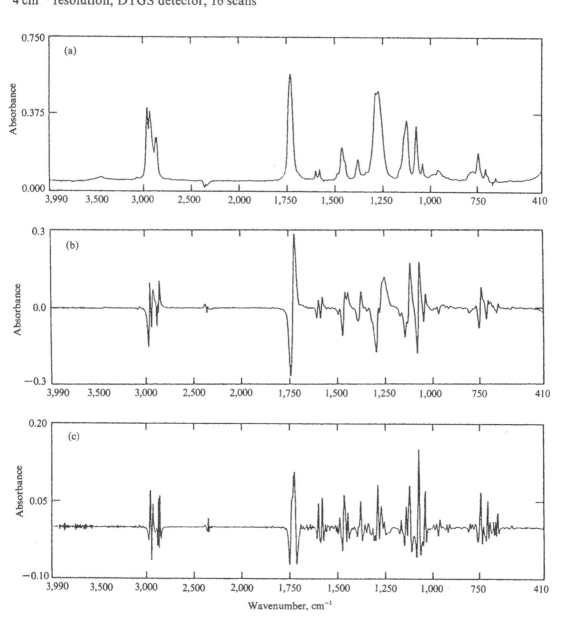

Wavenumber, cm^{-1}

11.2B | Study of Isomer Using Derivative Spectroscopy

The spectrum in the chart is the absorption spectrum of di-*sec*-butyl phthalate (a) and its second derivative spectrum calculated in absorbance units, followed by conversion into $\%T$ units (b). Although many peaks are further split into several lines, for example, the triplet band of the CH stretching vibration is split into five lines ($3,000 - 2,700$ cm^{-1}), the most interesting is the splitting of the carbonyl band at $1,740$ cm^{-1} into three components.   This suggests that the carbonyl band is composed of contributions form three different kinds of carbonyl compounds.   Mayo *et al.*[1] explained the spectral features in terms of three geometric isomers of di-*sec*-butyl phthalate as shown in Fig. 11.5.

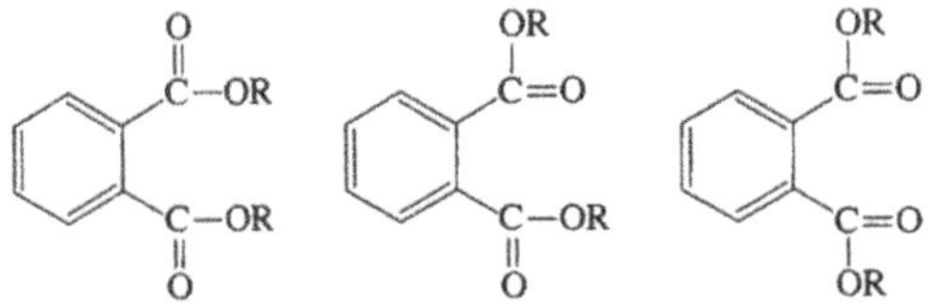

Fig. 11.5   Three isomers of di-*sec*-butyl-*ortho*-phthalate.

1.   R.W. Hannah, and D. Mayo, in *Infrared Spectroscopy*, Bowdoin College, Brunswick, Maine (1990).

Chart 11.2B

<Experimental conditions>

4 cm⁻¹ resolution, DTGS detector, 16 scans, second derivative obtained by Savitzky-Golay method using fewer smoothing points than used in the preceding section

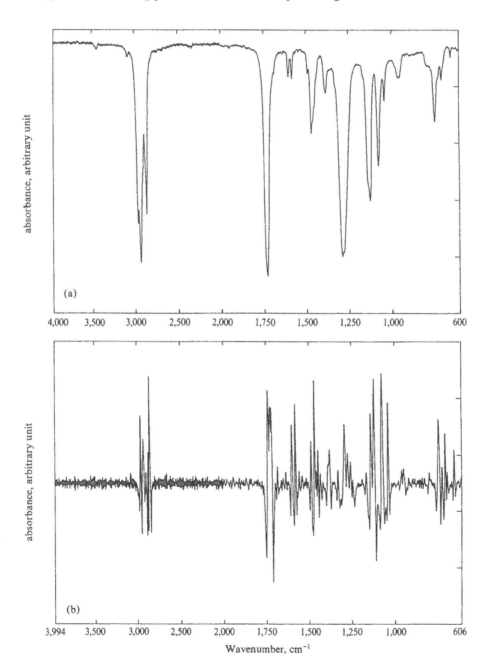

11.3   Simulation with Curve-Fitting

When a poorly resolved spectral band is observed for quantitative study, it may be desirable to separate the spectral features, into a sum of component bands, all of which have a different peak position, line shape, linewidth, and peak intensity.   Simulation techniques are used to perform such a separation.   The purpose of the simulation is to elucidate the peak positions and intensities of such component bands compared with the used peak height determinations done with an assumed baseline and peak position.

Based on theoretical discussions on the nature of linewidths of IR spectra, it was reported that the shape function of IR absorption bands become Lorentzian ($y = a^2 / x^2 + b^2$) or Gaussian ($y = a^2 \cdot \exp(-b^2 x^2)$) in certain cases.   Thus, Lorentzian, Gaussian, and/or a mixture of these two functions have been used to describe the shape of each component band.   The assumption in using these function is that the line shape of each component is symmetric.

An actual example of the simulation method is presented here, using the IR spectrum (a) of a drawn (10%) S-carbomethoxykeratin (SCMK) film described in Section 8C.   As given in the table in Section 8B, the amide-II band of the protein is composed of several peaks.   Thus, the position of each peak corresponding to random coil, α-helix, and β-pleat was first determined using a deconvolution technique.   Then, a trial-and-error simulation curve fitting method using the ratio of linewidth to intensity as a parameter was employed until the best fit (b) to the observed spectrum was achieved.

The peak positions and peak intensities which resulted in the best results are listed in Table 11.2. In the simulation, the mixture of Lorentzian function (spectrum (c) in the chart) and Gaussian function (spectrum (d) with a 9 : 1 ratio was used for *all* of the components because it gave the best fit to the observed spectrum.   The same linewidth was assumed for *all* of the components, because all of the component bands belonged to the same chemical structure but had different conformations.   The simulation was performed on all for the experimental data shown in Section 8C.

The simulation technique has been used to explain many important events in IR spectroscopy despite the tedious trial-and-error method, and assumptions such as a common line shape function and sometimes even a common linewidth as described in this scetion.   The simulation curve fitting method generally has quantitative applications and is complementary to deconvolution or derivative techniques.   A chemometric method such as multiple linear regression or partial least square generally have quantitative applications.   Therefore, the present authors wish to point out that the importance of the simulation technique has been decreasing rapidly since the development of chemometric methods.   Chemometric methods are described in a later section.

TABLE 11.2   Input data to simulate amide-II band of SCMK

Second-order structure		position	intensity
random coil		1536 cm⁻¹	0.26 abs.
α-helix	parallel absorption	1,516	0.61
α-helix	perpendicular absorption	1,540	0.79
β-sheet	parallel absorption	1,530	0.11
β-sheet	perpendicular absorption	1,550	0.27

Chart 11.3

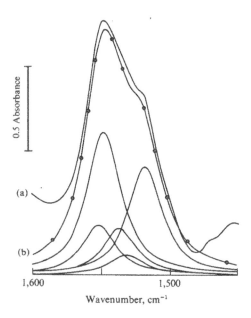

(a)

(b)

1,600   1,500

Wavenumber, cm⁻¹

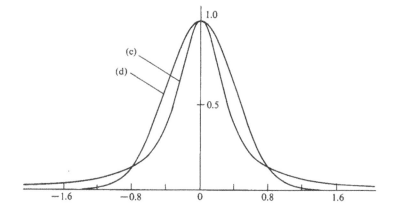

11.4   Kramers-Kronig Analysis

True specular reflection spectra have derivative-like spectral patterns and it is generally not advantageous to use them for qualitative or quantitative purposes. However, in some circumstances, the easiest way to obtain an IR spectrum from a sample is through a specular reflection measurement. Fortunately, as explained in Chapter 2, a reflection spectrum can be transformed into an absorption spectrum through the Kramers-Kronig integration. In this section, the application of K-K analysis is explained.

11.4A	Assignment of Plastics

We were interested in identifying a plastic pellet ($\sim 3 \times 3 \times 3$ mm). Since the surface of the plastic was smooth and flat, the specular reflection spectrum was expected to be free from any distortion due to diffuse reflection effects. Thus, the specular reflection spectrum of the plastic pellet was observed through a 2 mm aperture using a fixed angle specular reflectance accessory. Spectra (a) and (b) in the chart are the observed specular reflection spectrum and the absorption spectrum obtained through Kramers-Kronig analysis. An intense band at 3,300 cm^{-1} and a characteristic pair at 1,650 and 1,550 cm^{-1} indentify the material a secondary amide, a polyamide in this case. The flow chart for the identification of the polymer shown in the appendix also guides us to polyamide.

However, there are a several kinds of polyamides with different chemical structures as well as crystallgraphic isomers, α-form, β-form, γ-I, and γ-II forms. Thus, classification of polyamide requires caution. According to Ogawa,[1] the following peaks are utilized to identify polyamides

(Polyamide-6)	1,465, 1,265, 960, 925 cm^{-1}
(Polyamide-66)	1,480, 1,280, 935 cm^{-1}
(Polyamide-610)	1,480, 1,245, 940 cm^{-1}
(Polyamide-11)	1,475, 940, 720 cm^{-1}

Peaks in spectrum (b) obtained through Kramers-Kronig analysis are seen at 1,643, 1,543, 1,475, 1,441, 1,421, 1,374, 1,280, 1,205, 1,185, 1,150, 1,043, and 935 cm^{-1}. Judging from Ogawa's table, Polyamide-66 or Polyamide-11 is the most probable. However, based on the existence of a peak at 1,280 cm^{-1} and characteristic absorption due to α-form of polyamide, we concluded the mateial is the α-form Polyamide-66. It will be clear to readers that a specular reflection spectrum from the smooth and flat surface can be utilized after the K-K analysis. Since there are a large number of polyamides and crystal forms that affect IR spectra, unambiguous assignment can be achieved when the monomers of the polyamide are identified from hydrolysis of the polyamide.

1. T. Ogawa, *Handbook for Polymer Analysis,* ed. Japan Society for Analytical Chemistry, p. 321, Asakura Shoten, Tokyo (1985).

Chart 11.4A

<Experimental conditions>
4 cm^{-1} resolution, DTGS detector, 32 scans

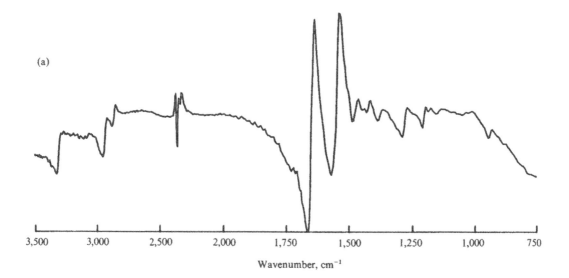

(a)

Wavenumber, cm^{-1}

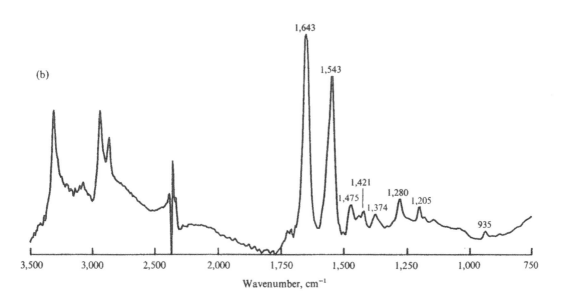

(b)

1,643

1,543

1,421

1,475

1,374

1,280

1,205

935

Wavenumber, cm^{-1}

11.4B   Surface Coating of Organic Lens

We analyzed two kinds of organic glass mateial most commonly used as lenses for eye glasses. One of them was coated with a fluorinated compound and the other was an untreated reference material.   Since the material is used as a lens, the coating layer is optically homogeneous and the material is known to have a distinct stratified bilayer structure.   As shown in Section 2.2, the ATR method may not be applied directly in the calculation of the difference spectrum of surface treatment material when the surface has a distinct bilayer structure.   It was demonstrated as shown in Fig. 11.6 that ATR difference spectroscopy does not show a peak at $1,280$ cm^{-1} as is characteristic of fluorocarbon type compounds (see Section 6.3E for such a spectrum) due to many oversubtracted and undersubtracted peaks of the organic lens material.

The specular reflection technique followed by KK analysis was attempted for this pair of samples.   Spectum (a) of the chart is a specular reflection spectrum of the coated organic lens and the absorption spectrum (b) obtained through Kramers-Kronig calculation.   Spectrum (c) is the absorption spectrum of the uncoated lens, obtained through KK analysis of the specular reflection spectrum.   A difference spectrum (d) between these spectra clearly shows the peak at $1,280$ cm^{-1}, although subtraction of the organic glass spectrum, especially the intense carbonyl group band at $1,740$ cm^{-1}, is not perfect.

As seen in this example, the ATR method involves difficulty in obtaining good difference spectra when the sample has a well–defined bilayer.   Specular reflection measurements followed by Kramers-Kronig calculation may be a better choice for such cases.

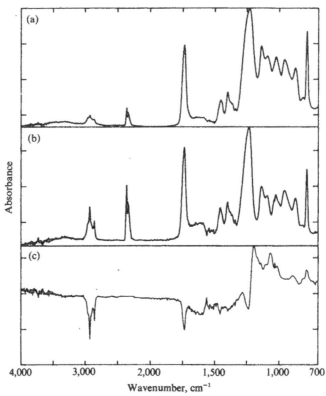

Fig. 11.6   ATR spectra of a surface coated organic lens (a) and an uncoated reference lens (b).   The difference spectrum between (a) and (b) is given in (c).

Chart 11.4B

<Experimental conditions>
4 cm^{-1} resolution, DTGS detector, 36 scans, fixed angle specular reflection accessory

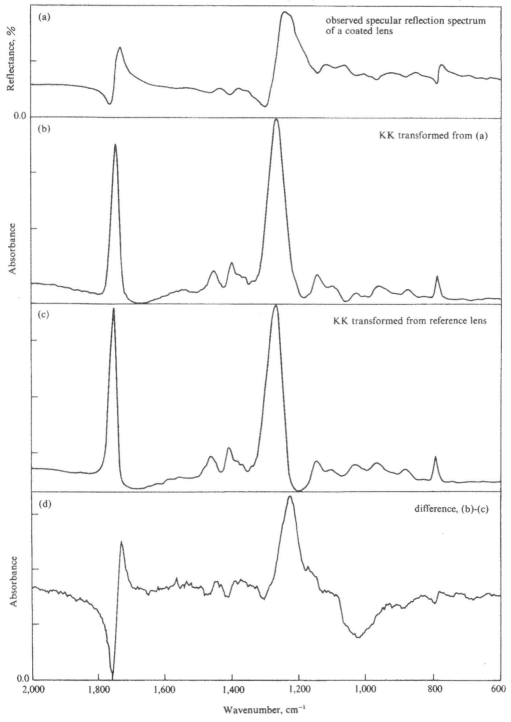

(a) observed specular reflection spectrum of a coated lens

(b) KK transformed from (a)

(c) KK transformed from reference lens

(d) difference, (b)-(c)

11.4C | Absorption Spectrum of Intractable Sample (β-SiC)

The sample thickness must be small when the absorption spectrum is measured for a sample which has strong absorptions.   For example, the absorption spectrum of NaCl can be obtained when the thickness is smaller than 1 μm.   However, it has been reported that the peak position shifts as the thickness of NaCl changes.[1]   A similar effect is seen in the case of Si_3N_4 formed on Si wafer.[2]

We measured the specular reflection spectrum of an optically polished plate of β-SiC, which is hard to prepare for other IR measurement techniques.   Spectrum (a) of the chart is the specular reflection spectrum.   As the frequency goes to *ca.* 1,000 cm^{-1} from the higher frequency, reflectance approaches to 0%R, because the refractive index of the sample approaches 1 around this frequency.   After this point, the reflectance changes rapidly to almost 95%R.   Then the reflectance decreases.   The absorption spectrum as obtained through Kramers-Kronig analysis is given in spectrum (b) in the chart.   The absorption band was found to be asymmetric and the peak position was 802 cm^{-1}.   When the sample is intractable, so that preparing a thin film is difficult, the Kramers-Kronig analysis of the specular reflection spectrum provides a fairly reasonable absorption spectrum.   Other samples suited for specular reflection measurements are smooth plates of glasses, ceramics, and plastics.   The reader may also recall the specular reflection measurement of the single crystal sample followed by the Kramers-Kronig integration, using polarized radiation (see Section 9.2A).

1. R.B. Barnes, M. Czerny, *Z. Phys.,* **72**, 447 (1931).
2. W.R. Knolle, D. L. Allara, *Appl. Sectrosc.,* **40**, 1046 (1986).

Chart 11.4C

<Experimental conditions>
4 cm^{-1} resolution, DTGS detector, 16 scans, specular reflection accessory (6.5° incident angle)

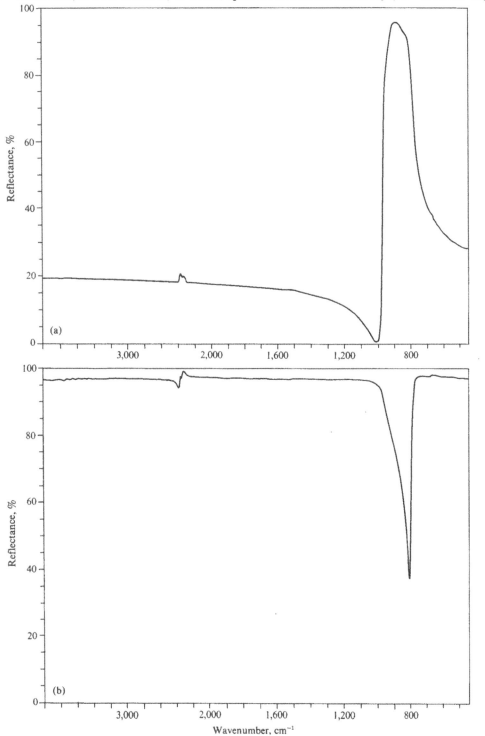

11.5   Chemometrics

In this section, we will review the current algorithms for quantitative and property analysis, since such computer software programs are available from FT-IR manufacturers and software houses.   Traditionally, a linear relationship between absorbance and product of concentration and sample thickness, the Lambert-Beer law, has been used in IR spectroscopy.

This relationship is normally written as $A = a \cdot b \cdot c + \varepsilon$, where a, b, and c are absorption coefficient, sample thickness, and concentration, respectively.   The term, ε, represents error. In this discussion, we rewrite the equation as, $A = BC + E$ where A, B, C, and E are absorbance, absorption coefficient (including sample thickness), concentration, and error, respectively. Since the computer can handle large numbers of data points, the first development in computerizing such quantitative analysis was achieved using full or selected ranges within the spectra of many standards, so that the Lambert-Beer law is given in matrix form and the calibration was performed to calculate B from A (observed spectra) and C (concentration) with a least-squres method.   This method is usually called the "Classical Least-Squares Method" (CLS) or "curve-fitting method."

Major progress in quantitative analysis was achieved when the multivariant statistical methods utilizing factor analysis became available to FT-IR users.   There are two major programs which are known by the general names PCR (Principal Component Regression) and PLS (Partial Least Squares).   In the factor analysis method, a series of IR spectra of the standards are again considered as the matrix, $A(n, m)$, each column being an IR spectrum in absorbance units and each row being a frequency.   Instead of solving for the product of concentration and coefficient matrices, the matrix, A, is transformed to the product of a loading spectrum, B, and score, T, and error, E, as

$$A = TB + E$$

Although the above equation appears to be equivalent to the above-mentioned curve fitting method, the meanings of these terms are different.   The factor analysis method employs a matrix diagonalization procedure to transform the A matrix to TB.   And, T is not concentration itself but concentrations will be obtained from the score, T, through a least-squares method.

The most important feature of the factor analysis method is that the method performs not only the quantitative analysis of the mixture in terms of component concentration, but can also provide the selected properties of the sample.   For example, an FT-IR system has been calibrated to determine the octane number of gasolines using a set of IR spectra taken from standard gasolines.   From the traditional spectroscopic view, there is no absorption peak directly related to the octane number.   However, since the chemical composition of the gasoline determines the octane number and the chemical composition is in essence related to the IR spectrum, the multivariant approach can select that part of the data set which is related to octane number as well as other physical properties.

In addition, the factor analysis method can reveal structures in a data set.   For example, the factor analysis method divides a data set into groups (see Section 8A), finds trends, and segregates outliers.   Another important feature is that the factor analysis can calculate pure component spectra from the mixture.   The reader will find it necessary to study the factor analysis methods, since these methods are becoming a standard data handling technique for anyone who works in the analytical field.[1]

1.   (a) D.M. Haaland and E.V. Thomas, *Anal. Chem.*, **60**, 1193 (1988), D.M. Haaland and E.V. Thomas, *Anal. Chem.*, **60**, 1202 (1988) and reference therein; (b) E.R. Malinowski, *Factor Analysis in Chemistry*, 2nd Edition, Wiley-Interscience, New York (1991); (c) P.M. Fredericks, J.B. Lee, P.R. Osborn and D.A. Swinkels, *Appl. Spectrosc.*, **39**, 303 (1985).

11.5A | Determination of Gasoline Octane Number [1]

The octane number of gasoline is used as an index to characterize the performance of the gasoline in the engine. Octane numbers of gasolines are usually determined based on ASTM D2699,[2] using a standard engine. However, Asker and Kokot[3] showed that octane number can be evaluated from near-infrared spectra of gasolines by a principal component analysis (PCA) followed by multiple linear regression (MLR). A model experiment is discribed here to demonstrate the usefulness of mid-IR spectra of gasolines analyzed using chemometrics to determine octane number.

Mid-IR spectra of seven samples with different octane numbers ranging from 80 to 100 were measured using a horizontal ATR (HATR) accessory. These spectra are shown in Fig. 11.7. As shown in Fig. 11.8, the PCA technique calibrates the FT-IR/HATR system as a device to determine gasoline octane numbers ($R^2 = 0.9705$).

In the chart, the property correlation spectrum for octane number (a), and principal component 1, PC1, (b), and principal component 2, PC2, (c) are shown for the $2,000 - 850$ cm^{-1} range. Each property correlation spectrum is indicative of a relationaship between the IR spectra and changes in property. Spectrum (a) shows strong positive correlation between octane number and concentration of aromatic components of gasoline whose absorption bands are seen at 1,605, 1,500, 1,100, and 1,000 cm^{-1}. It also shows a strong negative correlation with methylene and methyl groups as shown by negative peaks at 1,460 and 1,380 cm^{-1}, respectively. Frequency regions showing both strong positive and negative correlations are the preferable regions for calculating properties while those showing poor correlation are not.

The present PCA/MLR method found two factors which would reproduce the origial data set.

1. M. Morimoto and E. Nishio, *Chemistry Express,* **7**, 505 (1992).
2. Annual Book of ASTM Standards, Petroleum Products, Lubricants and Fossil Fuels (ASTM, Philadelphia), vol. 05. 04, pp. 13, 21, 26 (1990).
3. N. Asker and S. Kokot, *Appl. Spectrosc.,* **45**, 1153 (1991).

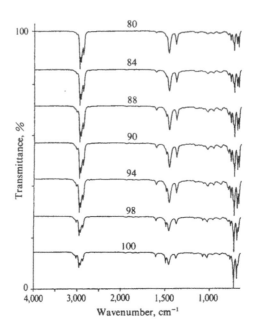

Fig. 11.7   ATR spectra of gasoline samples with different octane numbers.   The number over each spectrum indicates octane number.

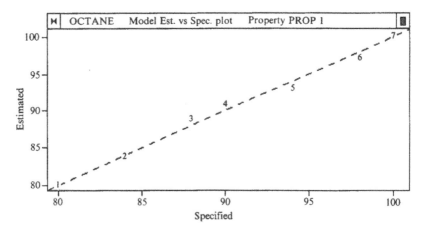

Fig. 11.8   Correlation between certified and estimated octane numbers of gasolines.

Chart 11.5A

<Experimental conditions>
4 cm^{-1} resolution, DTGS detector, 36 scans coadded, horizontal ATR accessory

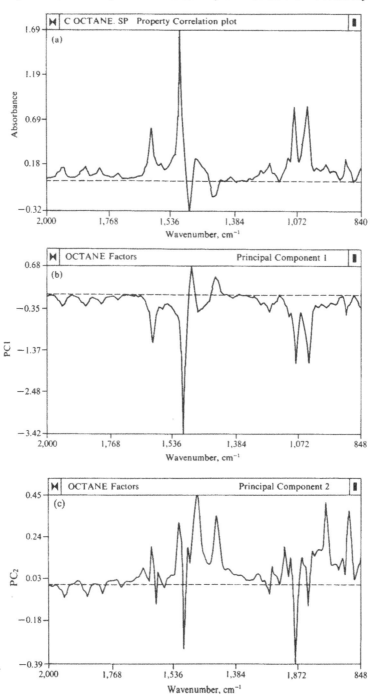

11.5B | Grouping and Trend Analysis by Factor Analysis*

A set of IR spectra taken during the repeatability test of a horizontal ATR accessory with an FT-IR system was analyzed using factor analysis software to find causes of error. In this experiment, an oil sample was poured into a trough on a ZnSe IRE, and IR spectra were taken repeatedly without any change in the system. Then a set of IR spectra were taken with the trough reloaded. The scores of factor 2 were plotted *versus* the scores of factor 1 and shown in chart (a). The plot is separated into two groups, one of which is a tight cluster (runs 1 to 10) of repeats and the other a large scatter (runs 11 to 19) of reloadings. As shown in the enlarged scale for the repeats (chart (b)), there is a trend in the scattering, shifting in a direction from run 1 to run 10, while no evident trend was seen for the scattering of points for reloadings.

Projection of factors 1 and 2 in the direction of the trend gives rise to a spectral feature which was found to coincide with the vapor spectrum of 1,2-dichloroethane as shown in the chart (c).

1,2-Dichloroethane was used as a solvent to wash the ZeSe IRE and it was discovered that the solvent was somehow trapped in the trough and gradually evaporated in the sample compartment, indicating that the drying process after washing had not been adequate. In addition, it was concluded that this problem was caused by a flaw in the trough design, which allowed the solvent to be trapped. This example shows how the factor analysis method can reveal information buried in large amounts of data.

* Data supplied by R. Aries, D. Lidiard and R. Spragg, *Spectroscopy,* **5**, 41 (1990).

Chart 11.5B

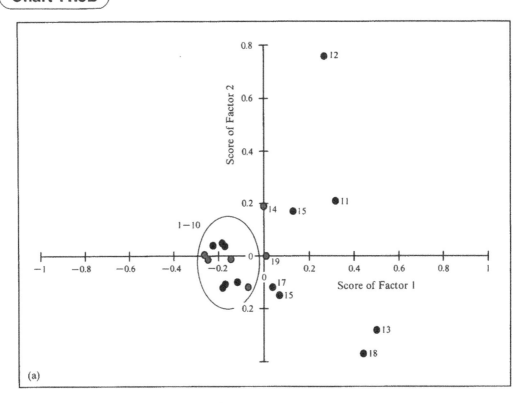

(a)

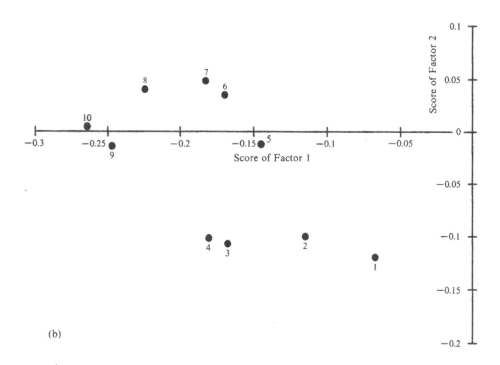

(b)

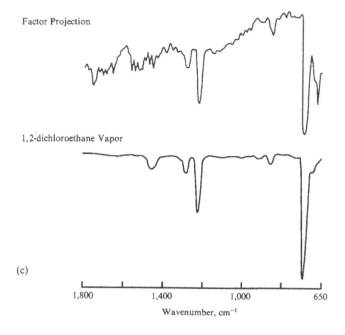

Factor Projection

1,2-dichloroethane Vapor

(c)

11.5C | Obtaining Pure Spectra from Mixture *

This section describes how to obtain pure spectrum of each component from a mixture,[1] using a simple model.   Let us explain how the scores of factor analysis are related to the input data. Fig. 11.9 is an unresolved peak in a gas chromatogram in which two different compounds are eluted as denoted.   A series of IR spectra were observed across the GC peak using a GC-FTIR system.   As shown in chart (a), only pure component A (isopropyl alcohol) was eluted during the initial stage (6 through 11) of the GC peak and pure component B (methyl propyl ketone) was eluted in the final stage (21 – 29) of the GC peak.   In other parts of the GC peak, spectra 12 through 20, mixtures of A and B were eluted.

Factor analysis of the input data showed that two factors were sufficient to represent the entire data set.   Plots of scores for factor 2 *versus* those for factor 1 are shown in chart (a).   Factor 1 resembles an average spectrum of the IR spectra of components A and B.   Factor 2, on the other hand, is the best fit to the residuals.   Thus, when only component A is eluted, scores for factor 1 are positive and scores of factor 2 are negative.   When only component A is eluted, the ratio of scores 1 and 2 is the same, although both scores increase as the concentration of component A increases.   Thus, the data on a straight line with direction of 310° shown in chart (a) correspond to pure component A.   On the other hand, the short straight line including a cluster along the 55° direction represents the time within the GC peak when only component B is eluted.   The intermediate part of the GC peak is on a smooth curve.

The IR spectrum of component A was calculated by the projection of two vectors, factor 1 and factor 2, along the 310° direction.   The spectrum of component B is obtained by the projection of those two vectors along the 55° direction.   Charts (b) and (c) illustrate the calculated spectra of components A and B (lower) and the observed spectra (upper).   As shown in this section, factor analysis method can be utilized to obtain pure spectra of components from mixtures. When there are more than two factors in the system, a multi-dimensional projection must be carried out.

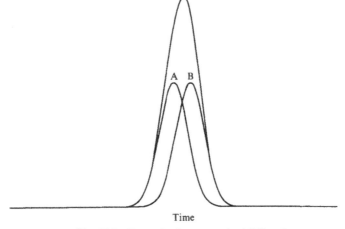

Time

Fig. 11.9   Example of an unresolved GC peak.

1.   For instance, P.J. Gemperline and J.C. Hamilton, "Factor Analysis of Spectro-Chromatographic Data" in: *Computer-Enhanced Analytical Spectroscopy,* Vol. 2, Chapter 2 (L.C. Meuzelaar ed.) Plenum, New York (1990).

* Data supplied by R. Aries, D. Lidiard and R. Spragg, *Spectroscopy,* 5, 41 (1990).

Chart 11.5C

(a)

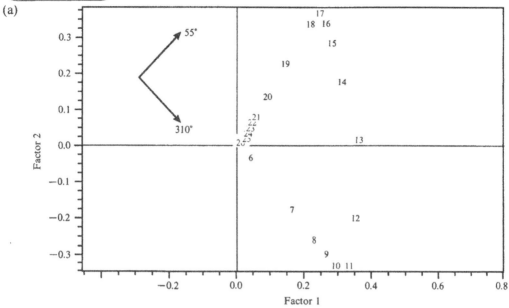

(b) Factor projection at 310°

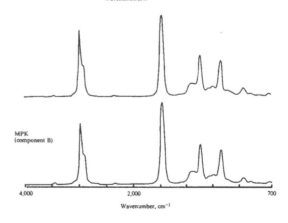

(c) Factor Projection at 55°

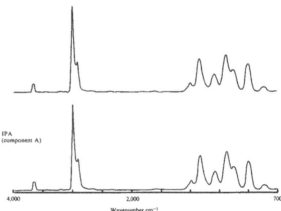

A Diagram to Classify Polymers by IR Spectra

A convenient method to classify natural and synthetic polymers into 21 groups using infrared spectra is shown in this appendix.   Starting from the left top box of the Fig. A, one can go step-by-step to one of the 21 groups by answering on the basis of the existence (downward) or non-existence (right) of an absorption band(s) described in the box.   The use of a computer search program will further pinpoint the assignment.

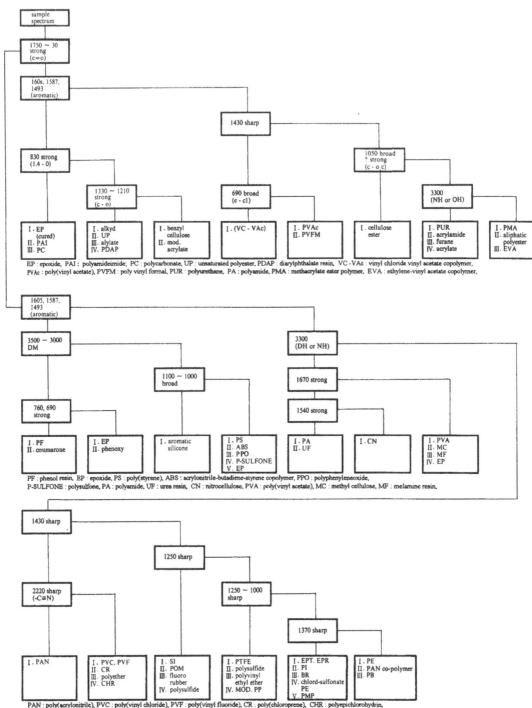

EP : epoxide, PAI : polyamideimide; PC : polycarbonate, UP : unsaturated polyester, PDAP : diarylphthalate resin, VC -VAc : vinyl chloride vinyl acetate copolymer,
PVAc : poly(vinyl acetate), PVFM : poly vinyl formal, PUR : polyurethane, PA : polyamide, PMA : methacrylate ester polymer, EVA : ethylene-vinyl acetate copolymer,

PF : phenol resin, EP : epoxide, PS : poly(styrene), ABS : acrylonitrile-butadiene-styrene copolymer, PPO : polyphenyleneoxide,
P-SULFONE : polysulfone, PA : polyamide, UF : urea resin, CN : nitrocellulose, PVA : poly(vinyl acetate), MC : methyl cellulose, MF : melamine resin,

PAN : poly(acrylonitrile), PVC : poly(vinyl chloride), PVF : poly(vinyl fluoride), CR : poly(chloroprene), CHR : polyepichlorohydrin,
SI : poly(methylsiloxane), POM : polyoxymethylene, PTFE : poly(tetrafluoroethylene), MOD. PP : modified poly(propylene),
EPT : ethylene-propylene terpolymer, EPR : ethylene-propylene rubber, PI : poly(isoprene), BR : butyl rubber, PMP : poly(4-methyl pentene-1),
PE : poly(ethylene), PAN : poly(acrylonitrile), PB : poly (butylene-1)

Fig. A

Index

Printed and bound by CPI Group (UK) Ltd, Croydon, CR0 4YY

23/10/2024

01778246-0014